Genetic Algorithms - Theory, Design and Programming

Edited by Yann-Henri Chemin

Published in London, United Kingdom

Genetic Algorithms - Theory, Design and Programming
http://dx.doi.org/10.5772/intechopen.107726
Edited by Yann-Henri Chemin

Contributors
Dharm Raj Singh, Duy Sang Nguyen, Rohollah Fallah Madvari, Cesar Guerra, Carlos A. Lara, Subhashis Mallick

First published in London, United Kingdom, 2024 by IntechOpen
IntechOpen is the global imprint of INTECHOPEN LIMITED, registered in England and Wales, registration number: 11086078, 5 Princes Gate Court, London, SW7 2QJ, United Kingdom
Printed in Croatia

British Library Cataloguing-in-Publication Data
A catalogue record for this book is available from the British Library

Additional hard and PDF copies can be obtained from orders@intechopen.com

Genetic Algorithms - Theory, Design and Programming
Edited by Yann-Henri Chemin
p. cm.

This title is part of the Artificial Intelligence Book Series, Volume 23
Topic: Evolutionary Computation
Series Editor: Andries Engelbrecht
Topic Editor: Sebastian Ventura

Print ISBN 978-1-83769-294-1
Online ISBN 978-1-83769-295-8
eBook (PDF) ISBN 978-1-83769-296-5
ISSN 2633-1403

For EU product safety concerns:
IN TECH d.o.o., Prolaz Marije Krucifikse Kozulić 3, 51000 Rijeka, Croatia,
info@intechopen.com or visit our website at intechopen.com.

IntechOpen Book Series
Artificial Intelligence
Volume 23

Aims and Scope of the Series

Artificial Intelligence (AI) is a rapidly developing multidisciplinary research area that aims to solve increasingly complex problems. In today's highly integrated world, AI promises to become a robust and powerful means for obtaining solutions to previously unsolvable problems. This Series is intended for researchers and students alike interested in this fascinating field and its many applications.

Meet the Series Editor

Andries Engelbrecht received the Masters and Ph.D. degrees in Computer Science from the University of Stellenbosch, South Africa, in 1994 and 1999 respectively. He is currently appointed as the Voigt Chair in Data Science in the Department of Industrial Engineering, with a joint appointment as Professor in the Computer Science Division, Stellenbosch University. Prior to his appointment at Stellenbosch University, he has been at the University of Pretoria, Department of Computer Science (1998-2018), where he was appointed as South Africa Research Chair in Artifical Intelligence (2007-2018), the head of the Department of Computer Science (2008-2017), and Director of the Institute for Big Data and Data Science (2017-2018). In addition to a number of research articles, he has written two books, Computational Intelligence: An Introduction and Fundamentals of Computational Swarm Intelligence.

Meet the Volume Editor

Dr. Yann-Henri Chemin has a Ph.D. in Algorithmic Remote Sensing Applications (Asian Institute of Technology, Th, 2006). He then graduated in Planetary Sciences with Astronomy in 2016 (Birkbeck, University College London, UK) and completed an eMBA in Security, Defence and Space in 2021 (University of Nice-Côte d'Azur, FR). He recently became Head of Technical Applications of Hyp4U, a French sovereign hyperspectral constellation. He previously worked on the hyperspectral analysis of the Apollo 12 landing site and its vicinity on the moon's surface. He also worked on the thermodynamic energy balance on the Earth's surface, as well as on the surface–atmosphere interactions of the moon Titan and on the mapping of craters on the dwarf planet called Ceres. Dr. Chemin's interests include challenges to settling on the moon and Mars, especially food creation, settlement construction, and in situ resource utilization.

Contents

Preface

Within the dynamic realm of computational intelligence, genetic algorithms (GAs) emerge as formidable tools for optimization, problem-solving, and machine learning. This book, *Genetic Algorithms – Theory, Design and Programming* presents a collection of scientific contributions that delve into the theoretical foundations of GAs while providing practical insights into their design and implementation.

GAs extend beyond the confines of academia, finding meaningful applications in societal and engineering domains. In societal contexts, from health care to urban planning, GAs optimize decision-making and resource allocation. In engineering applications, these algorithms revolutionize design processes, contribute to manufacturing optimization, and shape the evolution of artificial intelligence systems. Real-world examples and case studies within this volume bridge theoretical insights with practical applications, offering a compendium that demonstrates the potential of GAs in diverse scientific disciplines.

A solid understanding of programming principles is crucial to comprehending the scientific contributions within this volume. A dedicated section of this book guides readers through the practical aspects of implementing GAs in various programming languages. From coding fundamental algorithms to optimizing performance and handling real-world datasets, this edition aims to empower researchers with the tools to translate theoretical knowledge into robust scientific applications.

I hope that this collection is not only a testament to the scientific advancements in GAs but also a valuable resource for researchers, academicians, and practitioners. Through these pages, we invite you to partake in the ever-evolving knowledge bridging foundational theory and cutting-edge applications, fostering a deeper appreciation for the scientific contributions that shape the ever-changing landscape of GAs.

Yann-Henri Chemin
Joint Research Centre,
European Commission,
Ispra, Italy

Section 1

Theory and Design

Chapter 1

Optimization Using Genetic Algorithms – Methodology with Examples from Seismic Waveform Inversion

Subhashis Mallick

Abstract

Genetic algorithms use the *survival of the fittest* analogy from evolution theory to make random walks in the multiparameter model-space and find the model or the suite of models that best-fit the observation. Due to nonlinear nature, runtimes of genetic algorithms exponentially increase with increasing model-space size. A diversity-preserved genetic algorithm where each member of the population is given a measure of diversity and the models are selected in preference to both their objective and diversity values, and scaling the objectives using a suitably chosen scaling function can expedite computation and reduce runtimes. Starting from an initial model and the model-space defined as search intervals around it and using a new sampling strategy of generating smoothly varying initial set of random models within the specified search intervals; the proposed diversity-preserved method converges rapidly and estimates reliable models. The methodology and implementation of this new genetic algorithm optimization is described using examples from the prestack seismic waveform inversion problems. In geophysics, this new method can be useful for subsurface characterization where well-control is sparse.

Keywords: global optimization, genetic algorithm, inversion, parameter estimation, diversity preservation, sampling strategy

1. Introduction

Genetic algorithm (GA) is a global optimization method based on the natural analogy from genetics and evolution theory [1, 2]. In geophysics GA has been used to solve a variety of single- and multi-objective inverse problems [3–17].

GA belongs to the class of *model-based optimizations* in which there are three distinct components: (1) model, (2) data, and (3) objective. It is also assumed that the model and the objective are related to one another via data and the underlying physics of the problem. The model or the decision space is usually denoted as **X**. In seismic inverse problems under an isotropic elastic assumption for example, the model consists of the vectors of the P- wave velocity (V_P), S-wave velocity (V_S), and density (ρ) for each subsurface depth (or time) sample and the model space is the entire feasible range of their variability. For an anisotropic elastic inversion, the model would be other anisotropic parameters in addition to V_P, V_S, and ρ. For electromagnetic (EM) and gravity inverse problems, the model would consist respectively of the electrical resistivities or densities. Mathematically, the model or decision space is thus defined as

$$X = [x_1, x_2, \cdots, x_N]^T. \tag{1}$$

In Eq. (1) the superscript T represents a transpose and each $x_i, i = 1, 2, \cdots, N$ represents one model vector and N is the total number of model vectors. Thus for isotropic elastic seismic inverse problem at a single location, $N = 3$ with $x_1 = [V_{P1}, V_{P2}, \cdots, V_{PM}]^T, x_2 = [V_{S1}, V_{S2}, \cdots, V_{SM}]^T$, $x_3 = [\rho_1, \rho_2, \cdots, \rho_M]^T$, where for any depth or time sample $j, j = 1, 2, \cdots, M$, V_{Pj}, V_{Sj}, ρ_j respectively denote the P-wave velocity, S-wave velocity, and density at that sample, and M is the total number of samples. The data space is defined as

$$D = [d_1, d_2, \cdots, d_J]^T, \tag{2}$$

where each $d_i, i = 1, 2, \cdots J$ is the vector representing the ith data being optimized. Finally, the objective space is defined as

$$Y = [y_1, y_2, \cdots, y_J]^T, \tag{3}$$

where each $y_i, i = 1, 2, \cdots, J$ is a scalar valued quantity representing the objective of the ith data. To compute the objective, it is assumed that there exists a unique mapping of the model space onto the objective space, i.e., $X \rightarrow Y$ via the underlying physics of the problem and the data space **D**. The aim of any model-based optimization is to find the model (or suite of models) in the model (decision) space that either minimize or maximize the objective.

The problem defined above is multi-parameter and multi-objective optimization where multiple parameters (Eq. 1) are simultaneously estimated from multiple datasets (Eq. 2) via optimizing multiple objectives (Eq. 3). Such multi-parameter and multi-objective optimizations have been previously used in geophysics to solve a variety of problems such as estimating anisotropic properties for mantle lithosphere from the splitting parameters of teleseismic S-waves and P-wave residual spheres [18], wave equation migration velocity inversion [19], estimating earthquake focal mechanisms [20], inverting multicomponent seismic and electromagnetic (EM) data [10–12, 14, 17], etc.

This chapter restricts to the discussion of the multi-parameter and single-objective optimization problem such that there are multiple parameters in the models space to be estimated (Eq. 1) using a single set of data and single objective, i.e., when $J = 1$ in Eqs. (2) and (3). Although the examples provided are from the seismic inversion problem, its extension to solving other optimization problems is straightforward.

2. Multi-parameter and single-objective optimization problem

Restricting to seismic inversion under isotropic assumption, the model consists of three parameters, (1) the longitudinal or P-wave velocity (V_P), (2) transverse or the S-wave velocity (V_S), and (3) density (ρ). Thus, for single-component isotropic elastic seismic inverse problems, the model and data can be redefined as

$$m = [V_P, V_S, \rho]^T, \tag{4}$$

and

$$d = [d_1, d_2, \cdots, d_N]^T. \tag{5}$$

In Eq. (4), the vectors $V_P = [V_{P1}, V_{P2}, \cdots, V_{PM}]^T$, $V_S = [V_{s1}, V_{S2}, \cdots, V_{SM}]^T$, and $\rho = [\rho_1, \rho_2, \ , \rho_M]^T$ respectively denote the P-wave velocity, S-wave velocity, and density for each depth (or time) sample $i; i = 1, 2, \cdots, M$, and the vector d in Eq. (5) is the input seismic data with N samples.

Having defined the model and data, a unique forward modeling operator $s = g(m)$ is then defined which allows computing the synthetic or predicted data vector $s = [s_1, s_2, \cdots, s_N]^T$ exactly at the same points as those of the data vector d of Eq. (5). This forward modeling operator varies depending upon the flavor of the inversion method. For post-stack or pre-stack amplitude-variation-with-angle (AVA)/elastic-impedance inversion [9, 20–24], $g(m)$ is the convolutional modeling at normal or nonnormal incidence angles [25]. For wave-equation based inversion such as the full waveform inversion (FWI) [26–39], $g(m)$ is the numerical solution of the elastic or acoustic wave equation using finite-difference or finite-element modeling. Finally, for prestack waveform inversion (PWI), which is a subset of FWI under the assumption of a locally horizontal (1D) stratification at each location [3–8, 13, 16, 40, 41], $g(m)$ is the analytical solution to the 1D wave equation [42, 43].

After defining the synthetic or predicted data $s = g(m)$, the misfit or error between the observed and synthetic data is defined as $e = |d - s|$, following which, the objective is defined as the sum of the squared errors

$$y = e^T e, \tag{6}$$

and the optimization can be cast as the minimization of the objective y. Alternatively, the objective can be either defined as $-e^T e$ [17] or as the normalized cross-correlation [3]

$$y = \frac{d^\dagger s + s^\dagger d}{d^\dagger d + s^\dagger s}, \tag{7}$$

and the optimization can be cast as the maximization of y. The superscript † in Eq. (7) denotes a complex conjugate transpose. It is assumed here that the real and synthetic data vectors **d** and **s** can be complex valued. However, when they are real-valued, $d^\dagger = d^T$ and $s^\dagger = s^T$. Also, the objective defined by Eqs. (6) and (7) represent a pure least-square optimization which is unstable for most practical problems. Therefore, additional regularization or damping terms are used in defining the objective. These damping terms stabilize the optimization by addressing the inherent noise that are present on the data.

Figure 1 illustrates single-objective and multi-parameter global optimization with reference to PWI. As shown in **Figure 1**, an initial model of V_P, V_S, and ρ, each as functions of depth is first estimated and the search windows around them are provided to define the model-space. In the presence of well-logs with borehole measurements of V_P, V_S, and ρ, they can be used for generating this initial model. In the absence of well-logs, initial V_P can be estimated from seismic data using velocity analysis or other advanced velocity estimation procedures such as traveltime tomography [44] and then established V_P-V_S [23, 24, 45, 46] and V_P-ρ relations [47] can be used to estimate initial V_S and ρ. Based on the local geology and the knowledge of the variability of V_P, V_S, and ρ, the model-space or search windows can then be defined around these initial models and a suite of models is generated within them. These suites of models can be generated in different ways by providing an appropriate probability distribution around the initial model. For the most unbiased case, the distribution could be uniform (boxcar) between the minimum and maximum values and the models are generated such that the value of V_{Pi}, V_{Si}, and ρ_i at any depth i can have any random value within their minimum and maximum search bounds specified for that depth with equal probability. Alternatively, a Gaussian distribution with the mean and standard deviation, dictated by the initial model and the search window can be defined for each depth, and depth-dependent models of V_P, V_S, and ρ can be randomly drawn from this distribution. After generating these suite of models, synthetic seismograms using each of these models are computed and matched with the observed seismic data. When there are adequate offset samples, prestack seismic data in time and offset domain can be decomposed into plane-waves in the intercept-time and ray-parameter (τ-p) domain [5]; synthetic seismograms can also be computed in τ-p domain and matched with the real seismic data using Eq. (7) to compute the objective for each model. In many practical situations however, the offset samples may not be adequate to perform an accurate plane-wave decomposition in the τ-p domain and it must be approximately done in the angle-domain instead [8]. Thus, using the P-wave velocity field of each of the models, the offset-domain seismic data are converted into the angle domain using an offset-to-angle transformation method

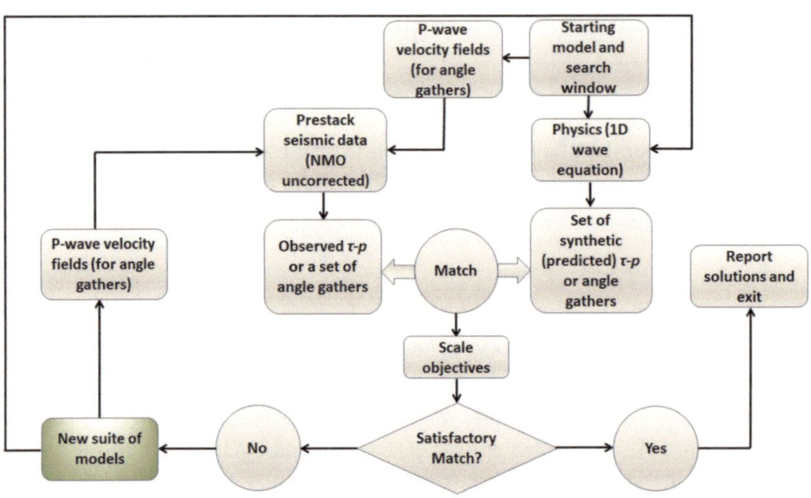

Figure 1.
Illustration of the single-objective and multi-parameter global optimization for prestack waveform inversion of seismic data.

[48–50]. Synthetic seismograms are also computed in angle domain and matched with the observed angle-gathers using Eq. (7) to compute the objective. Following the objective evaluation, they are scaled using a suitably chosen scaling function. Objective scaling, in principle, is equivalent to objective regularization and will be discussed in detail later in relation to genetic algorithms. Following scaling, the models generated so far are checked to verify if convergence is achieved. If yes, the method reports the solutions and exits. If not, the models are iteratively modified until a point is reached when one or more of the convergence criteria are satisfied (see **Figure 1**).

The way the models are modified in global optimization (the box shown with a different color in **Figure 1**), depends upon the flavor of global optimization being used. In simulated annealing (SA) for example [4, 6], the analogy of the fundamental physics of crystallization from a melt via slow and steady cooling is used to generate new models at each iteration. In Genetic algorithm (GA) on the other hand [1–3, 7, 8], the analogy from the natural selection and survival of the fittest of the evolution theory is used. In Particle swarm optimization (PSO), the models are modified using the analogy from the social behavior of a flock of birds or a school of fish [51, 52].

In theory, global optimizations do not depend upon the choice of the initial model like the local, gradient-based optimizations do. The only requirement for global methods is that the model-space provided as search bounds or prior probability distributions, is wide enough to contain the true model. In seismology, these global methods could be useful to characterize the subsurface in new exploration areas where well-control is sparse or unavailable. Even in matured areas with adequate well control, if production data are to be integrated with time-lapse seismic data through history matching via iterative reservoir simulations, geomechanical modeling, and seismic inversion; existing wells before production may not represent the true subsurface model and a global optimization would be useful to predict dynamic reservoir properties. Superiority of the GA over a linearized local inversion in predicting the ocean salinity and temperature from the water column reflections with no prior information has been clearly demonstrated by Padhi et. al. [41]. Yet, uses of global methods in seismic inversion is still limited. The primary reason for this is the fact that all global methods are nonlinear optimizations and computationally challenging for handling moderate to large sized seismic inverse problems. The advantage of using a global method is they can find a reasonable subsurface model even when the initial starting model is far from the true model. However, to find the true (global) model starting from a faraway initial guess requires (1) defining sufficiently large search bounds within the model-space, (2) generating a very large number of models within these bounds, and (3) iterating these models many times such that the model-space is adequately sampled (**Figure 1**). And the runtime for all global optimizations exponentially increase with increasing model-space size (i.e., with increasing search bounds) and increasing number of models to be iterated within this model-space [12].

Restricting specifically to GA, we will now elaborate the steps shown in **Figure 1** with particular emphasis on seismic inversion. As shown in **Figure 1**, the first step for GA is to get an estimate of the initial model and define the model-space. This initial model can be generated from the well-logs. But when well-logs are present, the initial model is close to the true model. Under such conditions, there are many compute-efficient local gradient-based methods to handle the inverse problem, and the use of a global optimization is unnecessary. The power of using any global optimization is in the situation where well-logs are unavailable, and the initial model must be differently estimated.

Figure 2 is a simple way to estimate the initial V_P via velocity analysis and normal moveout (NMO) correction. In this method, the input seismic data (**Figure 2a**) are interactively NMO corrected using different V_P fields until a time-varying V_P field, shown in **Figure 2b** is obtained that optimally flattens (NMO-corrects) the input gather (**Figure 2c**). Once the V_P field is estimated directly from data using the procedure shown in **Figure 2** or using other sophisticated method like tomography, initial V_S and ρ can be estimated using established V_P versus V_S and V_P versus ρ relations. To keep things simple, V_P-V_S relation, given as $V_S = \frac{V_P}{2}$ and the Gardner's relation [47] given as $\rho = \alpha V_P^{\beta}$ can be used to estimate the initial V_S and ρ. The V_P-V_S relation $V_S = \frac{V_P}{2}$ is commonly used in AVA/elastic-impedance inversion [23, 24, 46]. In Gardner's relation, $\beta = 0.25$ and α is 0.23 or 0.31, respectively for V_P in ft/s and m/s for most Gulf of Mexico sedimentary rocks.

Cyan curves in **Figure 3** represent the initial model for inverting the prestack seismic data shown in **Figure 2a. Figure 3a** is the initial V_P, which is same as the V_P shown in **Figure 2b** after time-to-depth conversion. Initial V_S and ρ, shown in **Figures 3b** and **c** were computed from the initial V_P of **Figure 3a** using the V_P-V_S and V_P-ρ relations as discussed above. The cyan curve of **Figure 3d** is the Poisson's ratio, computed from the initial V_P and V_S of **Figure 3a** and **b** using the formula for the Poisson's ratio, given as [53]

$$\nu = \frac{1 - 2\left(\frac{V_S}{V_P}\right)^2}{2\left\{1 - \left(\frac{V_S}{V_P}\right)^2\right\}}. \tag{8}$$

Note that because the initial V_S is computed from the initial V_P by setting $V_s = \frac{V_P}{2}$, per Eq. (8), the initial Poisson's ratio, shown in **Figure 3d** is constant $\left(\frac{1}{3}\right)$. The location

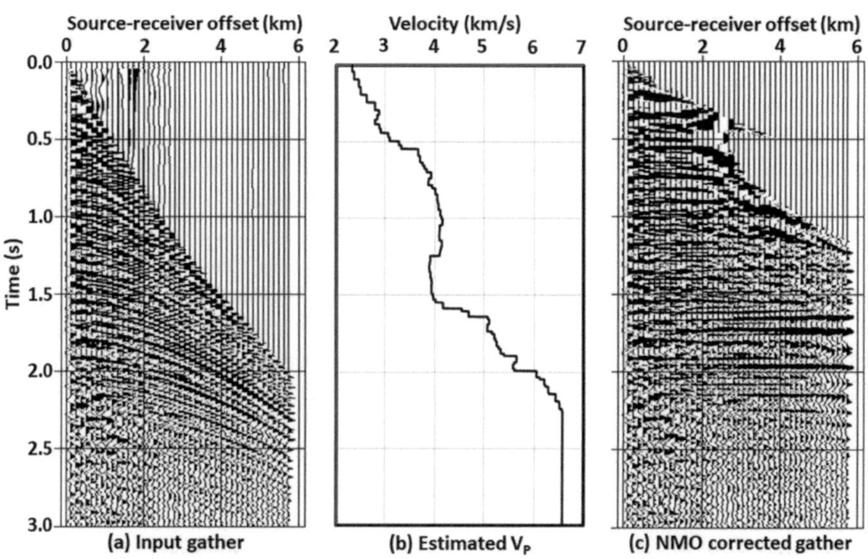

Figure 2.
Estimation of initial V_P from prestack seismic data. (a) Input seismic gather. (b) Estimated V_P. (3) Seismic gather after NMO correction using the estimated V_P.

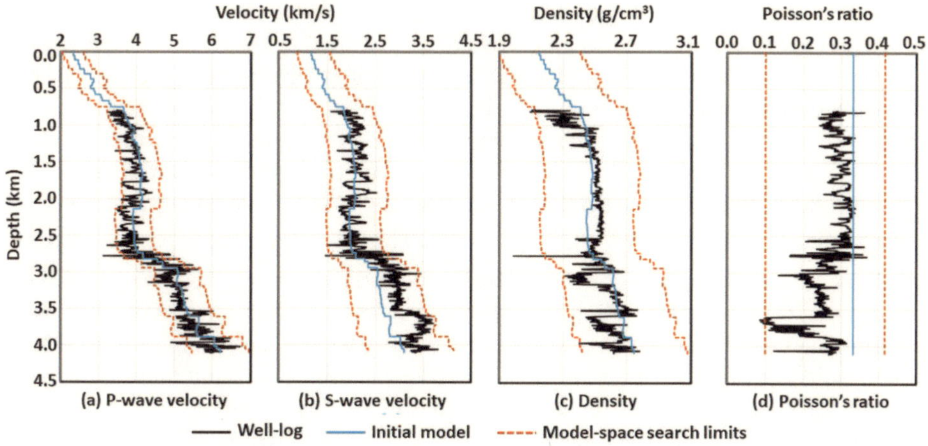

Figure 3.
The well-log model (black), initial model (cyan), and the model space search limits (dashed red) for (a) V_P, (b) V_S, (c) ρ, and (d) Poisson's ratio.

of the seismic data shown in **Figure 2** coincides with a well location and the well-log V_P, V_S, ρ, and the computed Poisson's ratio are shown as black curves in **Figures 3a–d**.

Besides the initial model, GA needs the model-space to be defined. The upper and lower bounds shown as dashed red curves on each panel of **Figure 3** are computed by varying (1) the initial V_P by $\pm 10\%$ (**Figure 3a**), (2) initial ρ by $\pm 15\%$ (**Figure 3c**), (3) initial Poisson's ratio from -70 to +25% (**Figure 3d**), and (4) by computing the upper and lower bounds of V_S (**Figure 3b**) from the initial V_P (**Figure 3a**) and the lower and upper bounds of the Poisson's ratio (**Figure 3d**). Note that the model space, defined by these upper and lower bounds (dashed red curves), are wide enough to span the range of variations in the true (well-log) model.

After estimating the initial model and the model-space (**Figure 3**), the next step in GA is generating the initial population of models. As mentioned earlier, these initial models can be generated by letting initial V_P, V_S (or Poisson's ratio), and ρ to randomly vary between their respective lower and upper bounds such that each of them can take any value between these bounds with equal probability. These models could be generated from the lower and upper bounds of V_P and ρ and the lower and upper bounds of either V_S or Poisson's ratio. Note that for single-component (i.e., for vertical or pressure response) seismic data considered here, the variations of reflection amplitudes are controlled by the variations of P-P reflection coefficients (R_{PP}) at subsurface interface boundaries as functions of the angle-of-incidence (θ). While the exact mathematical expression of R_{PP} is complex, linearized approximations suggest that $R_{PP}(\theta)$ is primarily controlled by (1) P-wave velocity contrast, (2) density contrast, and (3) V_S-V_P ratio (V_S/V_P) or Poisson's ratio contrast [53–59]. For both AVA waveform inversion and PWI, Mallick [7, 8] found that parameterizing the random models with V_P, Poisson's ratio and ρ, and then computing V_S from V_P and Poisson's ratio provides more stable inversion than parameterizing them directly in V_P, V_S, and ρ. From Eq. (8), it is straightforward to show that V_S can be obtained from V_P and Poisson's ratio (ν) as

$$V_S = V_P \sqrt{\frac{1-2\nu}{2(1-\nu)}}. \tag{9}$$

Thus, in the applications shown here, V_P, ν, and ρ were randomly generated and then Eq. (9) was used to calculate V_S. Also note that for 1D seismic inverse problems, layer thickness is also necessary. However, instead of layer thickness to be treated as the model parameter to be estimated, it is preferable to calculate them using a user-defined wavelength fraction at the dominant seismic frequency. Therefore, from the V_P value V_{Pi} for any layer i, the thickness for the layer i can be computed as $t_i = g\lambda_i$ where g is the user-defined wavelength fraction and λ_i is the wavelength at the dominant seismic frequency f_{dom}, i.e., $\lambda_i = \frac{V_{Pi}}{f_{dom}}$. Thus, the layer thickness also varies with the variation of V_P but controlled by the dominant seismic frequency and the user-defined wavelength fraction. For seismic inverse problems, it has been shown that allowing layer thickness to randomly vary along with V_P, V_S, and ρ is unstable and stable results are obtained by fixing them at the observed reflection boundaries in time domain [3, 4, 60]. Thus, in some old GA and SA implementations [3, 4, 60], the layers were of fixed time-thickness. Later Mallick [8] obtained stable results by resolving models at a given fraction of the dominant wavelength. More recently, a trans-dimensional seismic inversion method [61–65] have been developed where the number of layers besides V_P, V_S or ν, and ρ is treated as additional model parameter and estimated. However, in the applications shown here, a fixed number of layers with thickness set as the user-defined wavelength fraction is used.

Figure 4 shows randomly generated V_P models along with the search window, initial model, and true (well-log) model. For any layer i, the P-wave velocity V_{Pi} for the layer is assumed to have any random value between the minimum and maximum search limits $(V_{Pi})_{min}$ and $(V_{Pi})_{max}$ for the layer. Generating them for all layers and for all other model parameters (i.e., for V_S/ν and ρ) provides one random model and repeating the same for many models gives the model population. In **Figure 4**, ten random V_P models are shown with one of the models in orange and the other nine in grey. Note that when a sufficiently large population of such random models are generated, they would span the entire model space.

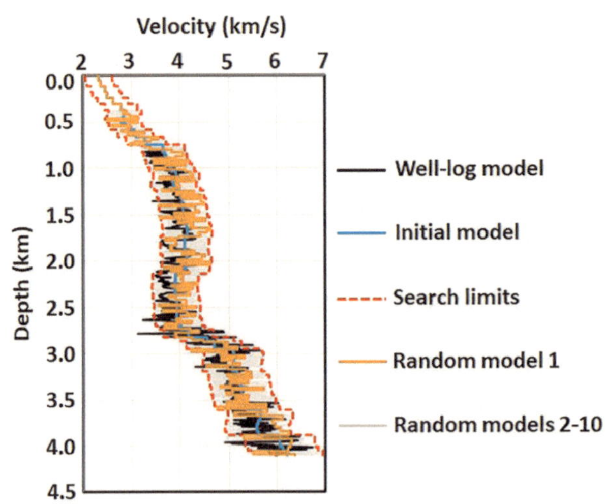

Figure 4.
Ten random V_P models generated within the specified search window. One of the models is shown in a different color than the others to show the variability of the model within the specified search limits. The well-log and initial V_P are also shown.

After generating the random population of models, synthetic data are computed for each model and matched with observation to compute the objective (**Figure 1**). However, before discussing objective computation, it is important to discuss one important GA-specific step known as the *parameter coding/decoding*. One popular version for GA works with coded parameters instead of the parameters themselves, and one good way for parameter coding is *binary coding*, illustrated in **Figure 5**. In binary coding, the model parameters (V_P, V_S or ν, and ρ) for each layer are coded as binary digits (bits). In **Figure 5** for example, 16 bits are assigned for V_P and 8 bits are assigned for V_S/ν and ρ. **Figure 5** shows the binary coded parameters for layer-1 only, but those for all layers are concatenated to form a single binary string to represent one model or one member of the population. In GA, this entire binary string representing one member is called a *chromosome* and each bit within the string is called a *gene*. The number of bits (genes) N_{bits} in a single model (chromosome) is thus given as $N_L\left(N_{VP} + N_{VS} + N_\rho\right)$ in which N_L is the number of layers and $N_{VP}, N_{VS},$ and N_ρ are the number of bits used to code V_P, V_S or ν, and ρ for each layer. Thus, in binary coding, a coin is tossed N_{bits} times with 50% probability for heads (1) and 50% for tails (0) and the results (i.e., zeros and ones) of the coin-toss are placed next to one another to represent one model (chromosome). Repeating this 2*N* times where *N* is an integer, produces a population of 2*N* models (**Figure 5**). For a single model parameter (V_P, V_S/ν, ρ) in each layer i, having all bits set to 0 (zero) corresponds to the minimum parameter value and having them all set 1 (one) corresponds to the maximum parameter value, specified by the search window for that layer (**Figure 4**). Thus, the bits with random zeros and ones from the coin-toss can be decoded for their actual value by linearly interpolating between these minimum and maximum values. For example, assume the following parameters for binary coding:

- For each layer, V_P, ν, and ρ are coded with 8 bits.
- Generated binary string for a single layer is given as 100101110011100111001001. And the search limits for the layer are 2500–5500 m/s for V_P, 0.1–0.42 for ν, and 2.0–2.5 g/cm^3 for ρ.

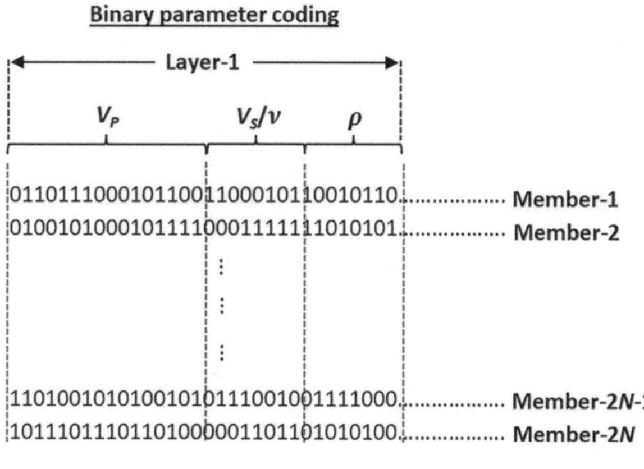

Figure 5.
Illustration of binary coding.

Under these conditions, V_P for the layer is represented by the first eight bits, i.e., 10010111, ν for the layer is the next eight bits- 00111001, and ρ is represented by the last eight bits- 11001001. Now the equivalent decimal integer d_{val} of the binary string 10010111 that V_P represents is $d_{val} = 1 \times 2^0 + 0 \times 2^1 + 0 \times 2^2 + 1 \times 2^3 + 0 \times 2^4 + 1 \times 2^5 + 1 \times 2^6 + 1 \times 2^7 = 1 + 8 + 32 + 64 + 128 = 233$. Assuming when all eight bits are zero (00000000) is equivalent to the minimum value $(V_P)_{min}$ of the search limit (2500 m/s) and when all of them are one (11111111) is equivalent to the maximum value $(V_P)_{max}$ of the search limit (5500 m/s), V_P for the layer is decoded as $V_P = (V_P)_{min} + (d_{val} - 1)\frac{(V_P)_{max} - (V_P)_{min}}{2^8 - 1} = 2500 + 232\frac{5500 - 2500}{255} = 5229$ m/s. Similarly, it can be shown that ν and ρ for the layer are respectively 0.295 and 2.29 g/cm^3. While such binary coding and decoding allows an easy way to search the model-space via crossover and mutation (see below), such a coding discreetly samples the model-space. For any model parameter P with minimum and maximum search limits P_{min} and P_{max}, the resolution of this discrete sampling is given as $\frac{P_{max} - P_{min}}{2^{N_B} - 1}$, where N_B is the number of bits used to code the parameter. Considering that the model space is, in fact, continuous, modern GA implementations are *real-coded*, which do not use any coding and directly work with real numbers. Thus, in real-coded GA, real values of the model parameters (V_P, V_S/ν, ρ) within the specified search window limits are randomly generated for each layer to represent one model or member in the population and do not require any decoding. Details of real parameter coding for GA can be found elsewhere [10–12, 66] and is not repeated here. Irrespective of whether the model parameters are coded as binary strings (**Figure 5**) or as real numbers, the next step in GA is to compute synthetic data using the underlying physics, match with the observation, and compute the objective.

Objective computation is followed by objective scaling or fitness scaling. In GA, objective is the primary driving mechanism for model-space search. However, if these objectives are directly used to drive search, the good models tend to dominate the poor ones, which is undesirable at early generations when none of the models are expected to be anywhere close to the true (global) optima, driving GA optimization to a local optimum [2]. In fitness scaling, the good models are scaled down and the poor models are scaled up and the degree of scaling up and down varies with generation. In GA, the original objectives computed using the normalized cross-correlation (Eq. 7 or equivalent) are called raw *fitness* and those after scaling are called *scaled fitness*. Although many ways for fitness scaling have been proposed, one simple and yet stable method is the linear scaling proposed by Goldberg (1989) [2]. In linear fitness scaling, the scaled fitness (f') is assumed to be linearly related to the raw fitness (f) as

$$f' = a + bf, \tag{10}$$

where a and b are constants, and are computed from the constraints

$$f'_{max} = S_C f_{avg}, \tag{11}$$

and

$$f'_{avg} = f_{avg}. \tag{12}$$

In Eq. (11), f'_{max} is the maximum value of the scaled fitness, f_{avg} is the average value of the raw fitness, S_C is the scaling constant, and f'_{avg} in Eq. (12) is the average scaled fitness, set equal to the average raw fitness f_{avg}. The scaling constant S_C is the user-defined parameter that controls the model selection and mathematically, it represents the expected number of copies of the best member to be selected for crossover and mutation (discussed below). In seismic inversion, setting $S_C = 0.8$ at the beginning and slowly increasing it to 1.8 at the final generation provides good results [7, 8]. As mentioned earlier, fitness scaling is like adding a regularization or damping term to the pure least square inverse problem.

Fitness scaling allows stability to GA optimization. In seismic inverse problems however, Sen and Stoffa (1992) [4] found that even after fitness scaling, GA may still fail to converge. This is because of *genetic drift*- a tendency for GA to cluster toward a single region in the model-space when finite population size is used [2]. One way to avoid genetic drift is to use a very large population size and propagate it to many generations. However, being a nonlinear method, the runtime for GA becomes prohibitively expensive with increasing population size and must be avoided. So, instead of using a large population, running several independent GA optimizations with small population sizes, and later combining them is one good way to handle genetic drift [4, 7, 17]. However, making several such GA runs is still compute-intense, and it is advisable to find ways to avoid genetic drift in a single rather than several GA runs. For multi-objective optimizations, one way to avoid genetic drift in a single run is to compute the *population diversity*, given as the normalized Euclidean distance of a member in the population from its nearest neighbors measured along all objective axes [67]. For single-objective optimization problem, diversity can be calculated using the pseudo-code shown in **Figure 6** in which the population size is given as N_P and the measured diversity is stored as a floating-point array *dist* of size N_P. Note that the diversity value calculated by the pseudo-code of **Figure 6** is normalized, which can have a value varying between 0 and 1. In addition, the higher the diversity value, the more diverse (i.e., more isolated from its neighbors) is the model. Thus, multiplying the raw fitness values of each model by their respective diversity values prior to fitness scaling would prefer more diverse models over the ones that are less diverse and add an additional control to avoid premature convergence of GA. Thus, in *diversity-preserved* GA, diversity is an additional regularization parameter for the objective besides fitness scaling.

Having discussed genetic drift and how to avoid clustering and premature convergence via objective-regularization, the next step in GA is *reproduction* or *tournament selection*. In this step, the models are selected in preference to their respective regularized (diversity-preserved and scaled) fitness values. Although many methods are

```
maximum_distance = 0.0;
loop over Np (ipop=0; ipop < Np; ipop++)
    dist[ipop] = 0.0;
    loop over Np (jpop=0; jpop < Np; jpop++)
        If (ipop ≠ jpop)
            dist[ipop] = dist[ipop] + |objective[ipop]-objective[jpop]|
        endif
    end Np loop (jpop=0; jpop < Np; jpop++)
    if (dist[ipop] > maximum_distance) dist[ipop] = maximum_distance;
end Np loop (ipop=0; ipop< Np; ipop++)
loop over Np (ipop=0; ipop < Np; ipop++)
    dist[ipop] = dist[ipop]/ maximum_distance;
end Np loop (ipop=0; ipop< Np; ipop++)
end
```

Figure 6.
The pseudo-code for diversity computation.

proposed for reproduction, the stochastic remainder selection without replacement [2] has been reported to work very well for seismic inversion problems [3, 7]. In this method the mathematically expected value E_i for each member i of the of the population are first calculated as

$$E_i = N_P \frac{f'_i}{\sum_{j=1}^{N_P} f'_j}. \tag{13}$$

In Eq. (13)N_P is the population size (number of models in the population) andf'_k is the diversity-preserved scaled fitness of the kth member in the population. After calculating E_i for all members, the integer part of E_i is used as the number of copies of the member i to be selected into a new population pool of the same size N_P. For example, if for a given member j, $E_j = 3.67$, then three copies of the jth member would be selected into the new population pool. Repeating this process for all members $j, j = 1, 2, , N_P$, would then partially fill up the new population pool, in which some of the original members would be present once, more than once, or not present at all, depending upon whether the integer part of their mathematically expected values are 1, greater than 1, or less than 1. Next, the rest of the new population is filled up as follows:

1. A member is randomly selected out of the entire original population and a coin is tossed with the probability of heads set to the fractional part of its expected value.

    a. Using the same example again, if the member j is randomly selected where $E_j = 3.67$, then the coin would be tossed with the probability of heads set to 0.67.

        i. If the outcome of the coin-toss is heads, then the one copy of the member j would be selected.

2. Step 1 above is repeated until the entire population is filled up, i.e., the selected number of members in the new population is N_P.

Reproduction generates a new population from the original (old) population, but the new population is simply a fitness and diversity preferred copy of the old population. To explore the model space and create a new generation of models from the old generation, the processes that GA uses are crossover, mutation, and elitism (update). Typically, crossover and mutation are combined into a single process, which is then followed by elitism to advance to the new generation. In the following, crossover, mutation, and elitism are explained using binary coding. Their extension for real-coded GA can be found in [66].

Figure 7 illustrates the binary-coded crossover. First, two members from the reproduced population are randomly selected and treated as parents. In **Figure** 7, they are called *Parent*-1 and *Parent*-2. Next, a crossover site within each parameter space (V_P, V_S/ν, ρ) for each layer is randomly selected (dashed purple lines in **Figure** 7). Finally, the bits (genes) on the left-hand side of the crossover sites of each parameter are swapped between the parents to produce two children, denoted as *Child*-1 and *Child*-2 in **Figure** 7. Crossover is performed with a crossover probability P_C. For each layer and each parameter, a coin is tossed with chances of heads set to P_C. If the

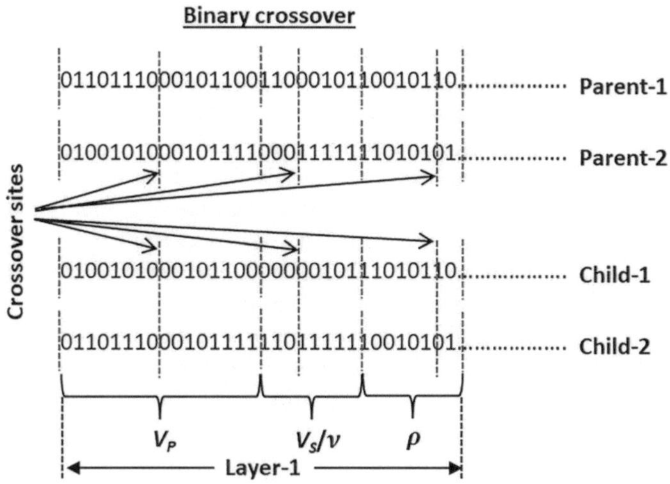

Figure 7.
Illustration of crossover.

outcome of the coin-toss is heads, then the crossover for the given layer and model parameter is performed, else the bits for the parents for the parameter are simply copied into the children (i.e., *Parent*-1 is copied into *Child*-1, and *Parent*-2 is copied into *Child*-2). Thus, crossover produces two new (child) members from the two original (parent) members. For seismic inversions, setting $P_C = 0.8$ at early generations and slowly reducing it to about 0.65 at the end produces good results [7, 8]. In mutation, for each bit (gene) in both child members (*Child*-1 and *Child*-2) is sequentially visited and a coin is tossed with the chances of heads set to the probability of mutation P_M. If the outcome of the coin-toss is heads, then the bit is changed, i.e., if it is 0 it is changed to 1 and if it is 1, it is changed to 0. A starting value of 0.15 and reducing it down to about 0.01 at the end is a good choice for P_M [7, 8].

After producing each pair of children from their parents via crossover and mutation, objectives for the children are computed. Next, in elitism the objectives of the children and their corresponding parents are compared and two members with the highest objective values are allowed to advance to the next generation. Like crossover and mutation, elitism is also performed with a probability of elitism P_E and a value of 0.5 at early generations and increasing it to about 0.9 at the end are good choices for P_E for seismic inversion problems [7, 8]. Crossover, mutation, and elitism produces two members of the next generation from the two members of the old generation and performing N such operations of crossover, mutation, and elitism for a population size $N_P = 2N$ would thus produce N_P members of the new (next) generation from N_P members of the old (previous) generation. Once the new generation of models are produced, they are advanced to another generation via reproduction, crossover, mutation, elitism, and the process is continued until a prespecified maximum number of generations- G_{max} is reached or some other stopping criteria (see below for details) is satisfied.

3. Practical implementation of GA

GA optimization is computationally challenging. However, the entire methodology can be parallelized in high-performance parallel computing environments, which, in turn, can lead to a compute-efficient nonlinear global seismic inversion method. Considering that a global method in place of a local, gradient-based method for seismic inversion can, in principle, allow estimating a reliable subsurface depth model of elastic properties, even when there is no prior (well) information, implementing an efficient global method like GA to solve seismic inverse problems would certainly be a big leap forward not only for oil and gas exploration, but also in solid earth geophysics, global seismology, and in the emerging fields like Carbon Capture, Utilization, and Storage (CCUS), Underground Hydrogen Storage in Porous media (UHSP) , characterization of the Enhanced Geothermal Systems (EGS), etc. However, to implement GA in practice, it is important to address a few important aspects. In above, GA is described as the process where an initial population models of size N_P, randomly generated within a user specified model-space (**Figure 4**) is propagated via reproduction, crossover, mutation, and elitism up to a specified number of generations G_{max}. While these are the fundamental GA steps, implementing just them, irrespective of how large the values of N_P and G_{max} are, may possibly work on synthetic data, but is most likely to fail when applied to real seismic data. To develop GA that would work on a wide variety of real data, fundamental issue that must be addressed is that the real data are always noisy, which must be efficiently handled. Any model-based optimization, whether local or global, is an iterative process which must have ways to stop iteration and exit such that the method does not suffer from unnecessary computational burden. Thus, after each iteration, these algorithms check if the method converged to a reasonable solution. If not, the iteration is continued, else the method reports solutions and exits (**Figure 1**). In dealing with noisy real data however, deciding whether to continue with iteration or to stop iterating is not one simple step as shown in **Figure 1**. Noise levels on real seismic data are known to widely vary, which are controlled by the geological factors, environmental factors, and many other factors directly related to how the data were acquired in the field and later processed in a computational facility. Even within a single area, there are often different noise levels in different parts. To handle noisy data, the stopping criteria for GA optimization must be implemented such that there is convergence check at different points within the algorithm, so that the data that are less noisy may converge early and exit and at the same time, more iterations are allowed for noisy data. Such multipoint convergence check, if correctly implemented, would not only allow additional computations when needed, but it would also avoid unnecessary computations on the data that are relatively less noisy.

Considering the above issues, **Figure 8** is the workflow for the prestack waveform inversion (PWI) using GA optimization. The workflow shown in **Figure 8** is complex, in which different parts or modules are shown with different colors and are outlined by dashed boxes, also of different colors. The first module (Module-1) is color coded in light green and outlined by the red dashed box. This module comprises the basic GA optimization steps. The second module (Module-2), color coded in peach and outlined by gray dashed box is the first convergence checkpoint. The third module (Module-3) in light blue and within purple dashed box is the second convergence checkpoint. Finally, the fourth module (Module-4), coded in yellow within green box is the output module where the method reports solutions and exits. The main user defined controlling parameters are (1) N_P– population size, (2) G_{max}– number of generations, (3) R_{max}– maximum number of repeats, (4) N_{iter}– maximum number of iterations, and (5) C_{min}– minimum correlation, i.e., the minimum value of objective (raw fitness value) to be achieved in the optimization process.

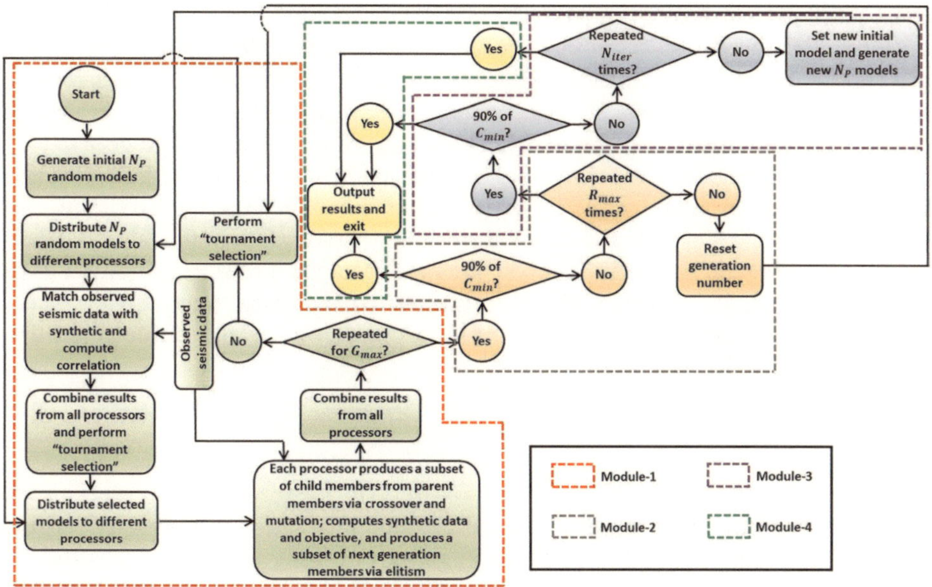

Figure 8.
Methodology for GA optimization with parallel implementation and multiple convergence checkpoints to solve seismic inverse problems.

At the beginning of Module-1 (modules with the red box), an initial random population of N_P models are generated and distributed to different processors. Different processors compute the synthetic data and objectives (raw fitness values) for the subset of the model population they each receive. The master node then collects results from all processors, computes diversity, and performs fitness scaling and reproduction (tournament selection). Next, the selected models are again distributed to different processors such that they each perform crossover, mutation, and elitism and produce new members of the next generation on a subset of the entire population. The size of the population subset each processor receives and produces new members is decided by the number of processors being used. For example, if $N_P = 100$ and 50 processors are used, then each processor would receive two members from the selected population as parents, produce two child members via crossover and mutation, and then two new members of the new generation via elitism. When the number of processors is less than 50, they each would then receive more than two parents and thus produce more than two new next generation members. If the number of assigned processors are more than 50, then only the first 50 processors would do the operation and the rest would sit idle; thus, for a population size of N_P, the maximum number of processors to be used by the method should be $\frac{N_P}{2}$, else the additional processors would not be used. In addition, N_P should be an even number so that $\frac{N_P}{2}$ is an integer. After generating new members within each processor, the master node receives them all and checks if the method progressed to G_{max} generations. If not, the master node continues to perform tournament selection using the new models and distributes the selected models to different processors for creating the models for the next generation. However, if G_{max} is reached, the methodology then goes into Module-2. In this module, whether the best value of the objective achieved at this point is at least $0.9C_{min}$ is first checked. If yes, the method goes straight into Module-4 (yellow) to report solutions and exit. Otherwise, the generation number is reset to 1 and the current model population is sent back to the green module (Module-1) to continue with another set of GA optimization for G_{max} generations, and this process is repeated for a maximum of R_{max} times with the chance of reporting solutions and exiting if $0.9C_{min}$is reached during any stage. After such R_{max} repeats, the method goes to Module-3. In this module, the convergence criterion of reaching the objective at least to $0.9C_{min}$ is first checked, and if not, a new initial model is set as the maximum likelihood model achieved at this point, a fresh new set of random models are generated within the model-space and sent to the top of Module-1 to start a fresh set of GA optimization with R_{max} repeats. Module-3 is repeated for a maximum of N_{iter} times with chances of being sent to Module-4 of reporting solutions and exiting at multiple points (see **Figure 8**).

4. Examples

Here, different PWI runs using GA optimization on a single real prestack seismic data are shown. Input seismic, initial model generated from velocity analysis and NMO correction and then using established V_P-V_S and V_P-ρ relations, and the search windows were used to define the model-space are shown in **Figures 2** and **3**. In all examples, GA with diversity preservation and linear fitness scaling was used. Fitness scaling constant (S_C), probabilities of crossover, mutation, and elitism (P_C, P_M, P_E) were all set to their recommended values as discussed earlier. Also, the models were parameterized as the P-wave velocity (V_P), Poisson's ratio (ν), and density (ρ).

Using the workflow of **Figure 8**, first two examples were run using $N_P = 80, G_{max} = 400, R_{max} = 5, N_{iter} = 7$, and $C_{min} = 2$. Note the value of C_{min} cannot exceed 1. In these two runs, it was deliberately set to 2 to ensure that the methodology goes through all seven iterations (N_{iter}), each with five repeats (R_{max}) of GA optimization using a population size (N_P) of 80 with 400 generations (G_{max}). In both examples, the search windows to define the model space were ±10% for V_P, -70% to +25% for ν, and ±15% for ρ around their initial values, which are shown as dashed red curves in **Figure 4**. Using 40 Intel Sandybridge/Ivybridge processors, runtimes for each were approximately 45 min.

In the first example, the initial random models were generated using the method shown in **Figure 4**, where the V_P, ν, and ρ for each layer were allowed to randomly vary within their respective model-space search limits, and the inversion result along with the initial model and the true (well-log) model are shown in **Figure 9**. The inverted model shown in red in this Figure is not the true inverted model, but a smoothed version of it, computed by taking a moving average of five samples (layers) across the entire depth range. The reason for showing a smoothed model instead of the actual model is because for unconstrained GA inversion where well-logs are not used at all, it is necessary to run multiple inversion passes. For the first pass, the initial model, generated from velocity analysis, V_P-V_S, and V_P-ρ relations is discretized at 0.5λ resolution where λ is the dominant wavelength as discussed earlier. In successive passes, a smoothed model from the previous pass is discretized at finer resolutions (0.25λ, 0.15λ, etc.) and used as the initial model. If the search window required for any given inversion pass is narrower than the one needed for the previous pass, then it can be inferred that the multi-pass inversion is moving toward convergence. Now, comparing the (smoothed) inverted model with the true (well-log) model shown in **Figure 9**, it can be readily verified that V_P is estimated reasonably well. However, the

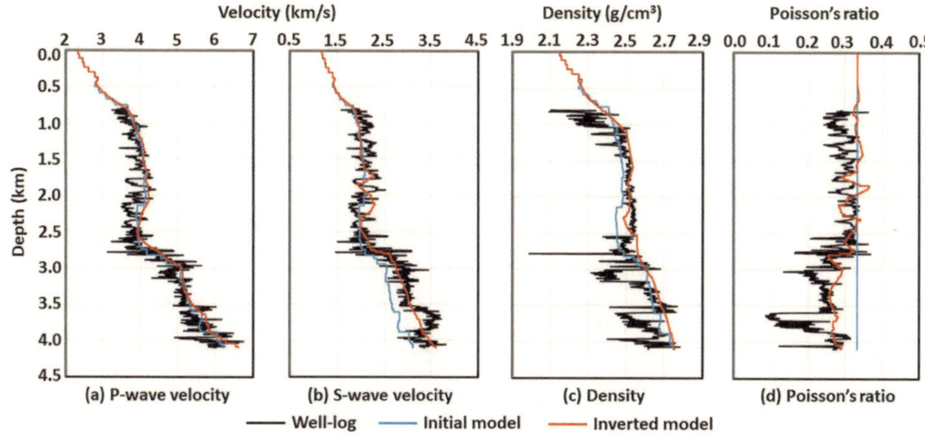

Figure 9.
PWI results using the diversity preserved GA optimization and random sampling method for initial model generation shown in **Figure** *4. (a) P-wave velocity, (b) S-wave velocity, (c) density, and (d) Poisson's ratio.*

estimates of V_S (or ν) and ρ are not as good, especially for depths greater than 3 km. Thus, by creating the initial random models using the sampling method of **Figure 4**, further refinements in the estimated inverted model is possible by making successive GA runs with the smoothed inverted model from the previous run set as the initial model for the next run with new search windows. Comparing the true (well-log) model with the initial model and the smoothed inverted model, shown in **Figure 9**, it can be readily seen that if the smoothed inverted model (the red curve in **Figure 9**) is used as the initial model for another inversion pass, much narrower search window than the first pass would be needed for GA to encompass all variations the true (well-log), and running a few such inversion passes would eventually find a model close to the true model. However, running such multiple inversion passes is not only compute-intense and cumbersome, but also impractical. Although well-logs were not used here to define the initial model, the results could still be compared with the well data and the inversion could be stopped when the estimated model is sufficiently close to the true well-log model. In the absence of well data, it is however difficult to come to a decision point of when the estimated model is sufficiently close to the true model so that the multiple inversion passes could be stopped. Thus, although the inversion result of **Figure 9** indicates that starting from a faraway initial model and using a wide model-space, GA optimization would, in theory, find the true model via successive inversion runs, the method is still difficult to apply in practice.

In contrast with the method for generating initial models shown in **Figure 4**, **Figure 10** shows a new way to generate them. Like **Figure 4**, **Figure 10** also shows ten random V_P models, generated between the minimum and maximum search bounds with model-1 in a different color from models 2–10. Note that the random V_P models of **Figure 10** span the entire search limit like the ones shown in **Figure 4**. These newly generated random models, however, vary vertically, but are much smoother laterally than the ones in **Figure 4**. Comparing the random models of **Figures 4** and **10**, the former could be regarded as the *laterally sampled* and later as the *vertically sampled* random models.

Figure 11 is the PWI result with GA optimization using the same parameters as the ones used in **Figure 9**, except the initial set of random models were generated via

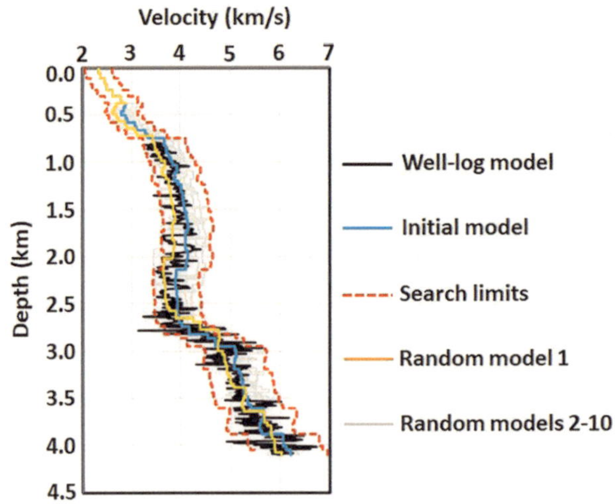

Figure 10.
*A new strategy of generating initial random models, demonstrated using V_P. Vertically varying and laterally smooth models are generated randomly between the minimum and maximum search limits. Like **Figure 4**, the search limits, well-log and initial V_P models, and model-1 in a different color than models 2–10 are shown.*

vertical sampling method of **Figure 10**. Note that unlike **Figure 9**, the smoothed inverted model of **Figure 11** is sufficiently close to the true model, and just a single inversion pass using this smoothed inverted model as the initial model should find the true model.

To verify, if the smoothed inverted model shown in **Figure 11**, would find the true model, another pass of PWI with GA optimization was run by using the model shown in red in **Figure 11** as the initial model and discretizing it at 0.25λ resolution. Search windows for V_P and ν were set to ±5% and that for ρ was set to ±2% and random models of population size $N_P = 40$ were generated using vertical sampling procedure of **Figure 10**. Other parameters used for this inversion were $G_{max} = 200, R_{max} = 3, N_{iter} = 3$, and $C_{min} = 1.0$. Like last two inversions, a diversity preserved GA optimization with linear

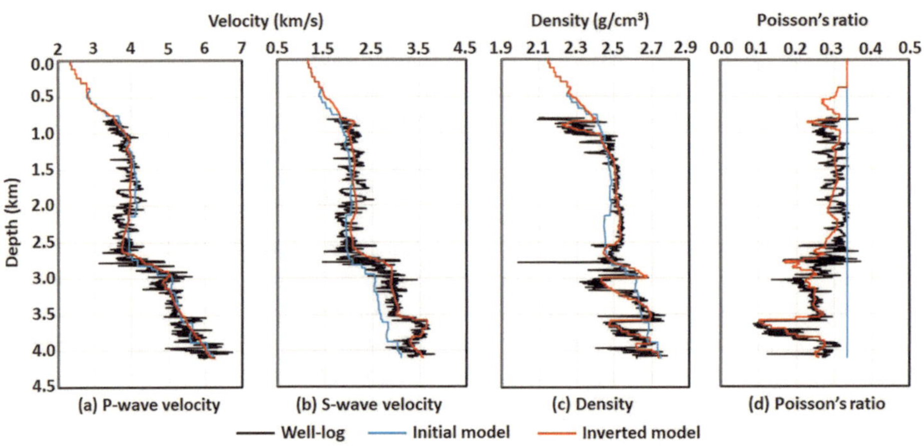

Figure 11.
*Same as **Figure 9** except when the initial set of random models were generated using the vertical sampling method of **Figure 10**.*

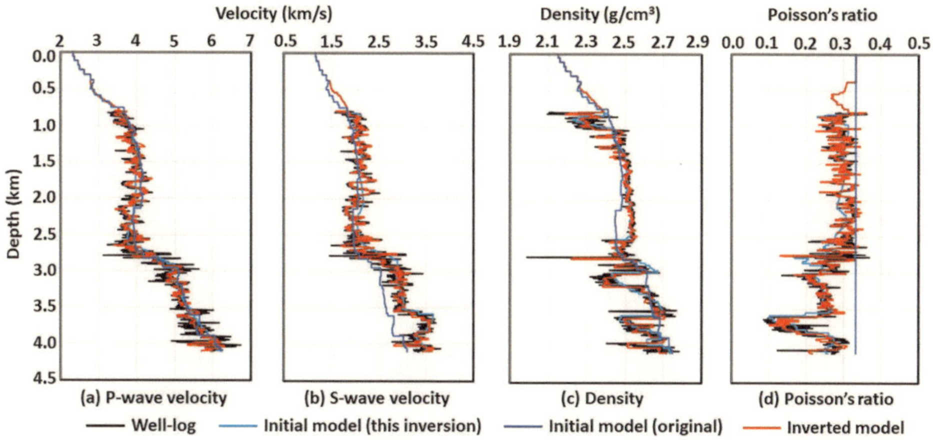

Figure 12.
Pass-2 inversion, using the smoothed inverted model of ***Figure 11*** *as the initial model. The original initial model is also shown in blue.*

fitness scaling was used with the same parameters as before. Using 20 Intel Sandybridge/Ivybridge processors, runtime for this inversion was 10 min and the results are shown in **Figure 12**.

Unlike **Figures 9** and **11**, the inverted model in **Figure 12** is the actual model estimated from inversion, not a smoothed version of it. This model is almost identical to the true (well-log) model, indicating that vertical sampling strategy for generating the initial set of random models is superior to lateral sampling and is the practical way for using GA optimization to solve seismic inverse problems when the initial model is far, and the model-space is large.

5. Discussion

Vertical sampling (**Figure 10**) instead of lateral sampling (**Figure 4**) is a new concept in the GA optimization for PWI. By starting from an initial model obtained directly from data and well-known V_P-V_S and V_P-ρ relations, using a large model-space defined as wide search windows for each model parameter, and generating vertically sampled initial set of random models (**Figure 10**), this newly proposed two-pass GA optimization can find the true (well-log) model with very good accuracy. By comparing lateral sampling (**Figure 4**) with the new vertical sampling (**Figure 10**) it may indicate the former sampling method encompasses the model space more uniformly than the latter. This is, however, untrue. In **Figures 4** and **10** only ten random models are shown, and when sufficiently large numbers of models are generated, both methods sample the model-space equally well.

Figure 13 compares a single random model of V_P, ρ, and ν generated from both standard (lateral) and newly proposed (vertical) sampling methods along with the initial model and the search limits. There is a fundamental difference between how the models are generated in these methods and how do they behave within the model space. In lateral sampling, values for each parameter and for each layer are independently generated in between their respective search bounds. Thus, V_P, ρ, and ν for each layer vary widely between their specified search limits (light green curves in

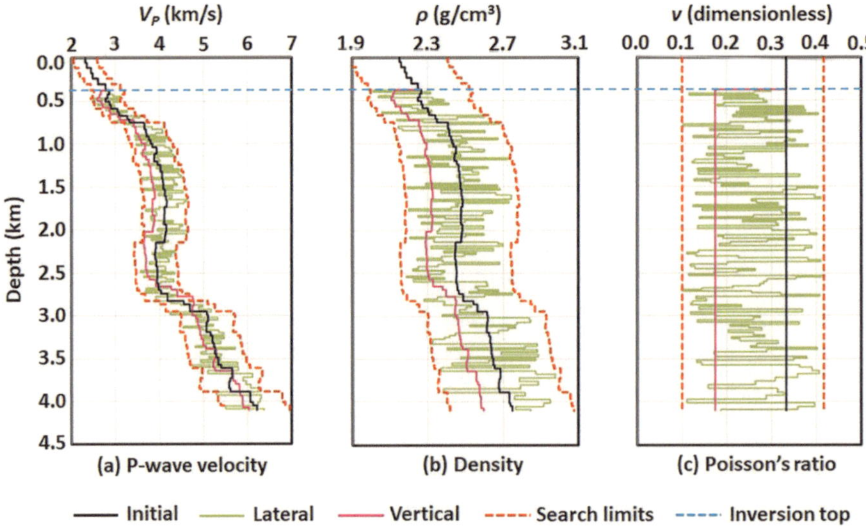

Figure 13.
Comparison between the lateral (light green) and vertical (magenta) sampling using a single random model. The search limits (dashed red) and the initial model (black) are also shown. In addition, the dashed line in cyan represents the top of the inversion window. (a) $\mathbf{V_P}$, *(b)* $\boldsymbol{\rho}$, *and (c)* $\boldsymbol{\nu}$.

Figure 13). Vertical sampling, shown as magenta curves in **Figure 13** on the other hand, uses a single random number for each model. Thus, a random number between 0 and 1 is generated and combined with the minimum and maximum search limits for each model parameter at each depth point or layer to define the model. For example, consider two specific layers in **Figure 13** at 1 and 2 km depths. The minimum and maximum limits for V_P at 1 km are approximately 3.3 and 4.2 km/s and those at 2 km are about 3.5 and 4.5 km/s. So, if the random number generated for a given model is 0.35, then the V_P at 1 km for the model would be set to $3.3 + 0.35 \times (4.2 - 3.3) =$ 3.615 km/s. Similarly, V_P at 2 km for the same model would be calculated as $3.5 +$ $0.35 \times (4.5 - 3.5) = 3.85$ km/s. Values for ρ and ν can also be computed in a similar fashion. Note that because the initial Poisson's ratio is constant, the random Poisson's ratio models generated by vertical sampling are also constant. Because a single random number defines one model, vertically sampled random models are much smoother than the laterally sampled ones. In addition, these vertically sampled random models follow the initial V_P, V_S (or ν) and ρ model trends. In the examples shown here, the initial V_P is derived from the velocity analysis and NMO correction of prestack seismic data, and the initial V_S (ν) and ρ were generated from this initial V_P and established V_P-V_S and V_P-ρ relations, which, in a broad sense, is the representative of the local geology that follows the regional compaction trend. Even for the case when well-logs are used as the initial model and then the initial random models are generated by vertical sampling, they would still be the representative of the geology. Thus, the vertically sampled random models could also be regarded as *geologically constrained* random models. Although random, their smooth and geologically constrained behavior is the reason for their ability to better sample the model-space. Starting from the combinations of V_P, V_S (or ν) and ρ models, consistent with the local geology, they tend to converge faster than those generated from lateral sampling. Finally, the dashed cyan line across all panels of **Figure 13** is the top of the inversion window. All model parameters above this line were regarded as overburden and kept unchanged.

Discussion here is restricted to the application of GA to solve seismic inverse problem a single location using the parallel implementation outlined in **Figure 8**. This procedure at single location is extendable to multiple location using the multi-level parallelization. The concept of such multi-level parallelization is to use multiple sets of the workflow of **Figure 8** to simultaneously invert different regions of the seismic data. Details of such multi-level parallelization is described elsewhere [13, 16] and therefore not repeated.

6. Conclusions

By introducing a new sampling strategy for generating initial random models, a GA optimization methodology is presented here. By avoiding genetic drift via diversity preservation and fitness scaling, the proposed method has been shown to work very well for large-sized model-spaces in solving seismic waveform inversion problems. The proposed method shown for a single location can be easily extended to multiple locations using a multi-level parallelization approach.

Author details

Subhashis Mallick
University of Wyoming, Laramie, WY, USA

*Address all correspondence to: smallick@uwyo.edu

References

[1] Holland JH. Adaptation in Natural and Artificial System. Ann Arbor, MI, USA: University of Michigan Press; 1975

[2] Goldberg DE. Genetic Algorithms in Search, Optimization and Machine Learning. Boston, MA, USA: Addison Wesley Publishing Company; 1989

[3] Stoffa PL, Sen MK. Nonlinear multiparameter optimization using genetic algorithms: Inversion of plane-wave seismograms. Geophysics. 1991;**56**: 1794-1810

[4] Sen MK, Stoffa PL. Rapid sampling of model space using genetic algorithms: Examples from seismic waveform inversion. Geophysical Journal International. 1992;**108**:281-292

[5] Sen MK, Stoffa PL. Global Optimization Methods in Geophysical Inversion. Amsterdam, Netherlands: Elsevier Science Publications; 1995

[6] Sen MK, Stoffa PL. Bayesian inference, Gibbs's sampler and uncertainty estimation in geophysical inversion. Geophysical Prospecting. 1996;**44**:313-350

[7] Mallick S. Model-based inversion of amplitude-variations-with-offset data using a genetic algorithm. Geophysics. 1995;**60**:939-954

[8] Mallick S. Case History: Some practical aspects of prestack waveform inversion using a genetic algorithm: An example from the east Texas Woodbine gas sand. Geophysics. 1999;**64**:326-336

[9] Du Z, MacGregor LM. Reservoir characterization from joint inversion of marine CSEM and seismic AVA data using Genetic Algorithms: a case study based on the Luva gas field. SEG Technical Program Expanded Abstracts. 2010;**80**:737-741

[10] Padhi A, Mallick S. Accurate estimation of density from the inversion of multicomponent prestack seismic waveform data using a non-dominated sorting genetic algorithm. The Leading Edge. 2013;**32**:94-98

[11] Padhi A, Mallick S. Multicomponent prestack seismic waveform inversion in transversely isotropic media using a non-dominated sorting genetic algorithm. Geophysical Journal International. 2014; **196**:1600-1618

[12] Li T, Mallick S. Multicomponent, multi-azimuth pre-stack seismic waveform inversion for azimuthally anisotropic media using a parallel and computationally efficient non-dominated sorting genetic algorithm. Geophysical Journal International. 2015;**200**:1136-1154

[13] Mallick S, Adhikari S. Amplitude-variation-with-offset and prestack waveform inversion: A direct comparison using a real data example from the Rock Springs Uplift, Wyoming, USA. Geophysics. 2015;**80**(2):B45-B59

[14] Li T, Mallick S, Tamimi N, Davis T. Inversion of wide-azimuth multicomponent vertical seismic profile data for anisotropic subsurface properties. SEG Technical Program Expanded Abstracts. 2016;**2016**:1252-1257

[15] Mazzotti A, Bienati N, Stucchi E, Tognarelli A, Aleardi M, Sajeva A. Two-grid genetic algorithm full-waveform inversion. The Leading Edge. 2016;**35**: 1068-1075

[16] Pafeng J, Mallick S, Sharma H. Prestack waveform inversion of three-

dimensional seismic data – An example from the Rock Springs Uplift, Wyoming, USA. Geophysics. 2017;**82**(1):B1-B12

[17] Ayani M, MacGregor L, Mallick S. Inversion of marine controlled source electromagnetic data using a parallel non-dominated sorting genetic algorithm. Geophysical Journal International. 2020;**220**:1066-1077

[18] Kozlovskaya E, Vecsey L, Plomerova J, Raita T. Joint inversion of multiple data types with the use of multiobjective optimization: problem formulation and application to the seismic anisotropy investigations. Geophysical Journal International. 2007; **171**:761-779

[19] Singh VP, Duquet B, Leger M, Schoenauer M. Automatic wave equation migration velocity inversion using multiobjective evolutionary algorithms. Geophysics. 2008;**73**:VE61-VE73

[20] Heyburn R, Fox B. Multi-objective analysis of body and surface waves from the Market Rasen (UK) earthquake. Geophysical Journal International. 2010; **181**:532-544

[21] Oldenburg DW, Scheuer T, Levy S. Recovery of acoustic impedance from reflection seismograms. Geophysics. 1983;**48**:1318-1337

[22] Oldenburg DW, Levy S, Stinson K. Root-means-square velocities and recovery of the acoustic impedance. Geophysics. 1984;**49**:1653-1663

[23] Connolly P. Elastic impedance. The Leading Edge. 1999;**18**:438-452

[24] Hampson DP, Russell BH, Bankhead B. Simultaneous inversion of pre-stack seismic data. SEG Technical Program Expanded Abstracts. 2005;**75**: 1633-1637

[25] Mallick S. Amplitude-variation-with-offset, elastic impedance, and wave-equation synthetics – A modeling study. Geophysics. 2007;**72**:C1-C7

[26] Pratt RG. Seismic waveform inversion in the frequency domain. Part 1: Theory and verification in a physical scale model. Geophysics. 1999;**64**: 888-901

[27] Vigh D, Starr EW. 3D prestack plane-wave, full-waveform inversion. Geophysics. 2008;**73**:VE135-VE144

[28] Plessix R-É. Three-dimensional frequency-domain full-waveform inversion with an iterative solver. Geophysics. 2009;**74**:WCC149-WCC157

[29] Lee KH, Kim HJ. Source-independent full-waveform inversion of seismic data. Geophysics. 2003;**68**:2010-2015

[30] Liu F, Guasch L, Morton SA, Warner M, Umpleby A, Meng Z, et al. 3-D time-domain full waveform inversion of a valhall obc dataset. SEG Technical Program Expanded Abstracts. 2012;**82**:1-5. DOI: 10.1190/segam2012-1105.1

[31] Guasch L, Warner M, Nangoo T, Morgan J, Umpleby A, Stekl I, et al. Elastic 3D full-waveform inversion. SEG Technical Program Expanded Abstracts. 2012;**82**:1-7. DOI: 10.1190/segam2012-1239.1

[32] Guitton A, Ayeni G, Diaz E. Constrained full-waveform inversion by model reparameterization. Geophysics. 2012;**77**:R117-R217

[33] Warner M, Ratcliffe A, Nangoo T, Morgan J, Umpleby A, Shah N, et al. Anisotropic 3D full-waveform inversion. Geophysics. 2013;**78**:R59-R80

[34] Bansal R, Krebs J, Routh P, Lee S, Anderson J, Baumstein A, et al.

Simultaneous-source full-wavefield inversion. The Leading Edge. 2013;**32**: 1100-1108

[35] Xue Z, Zhu H, Fomel S. Full-waveform inversion using seislet regularization. Geophysics. 2017;**82**: A43-A49

[36] Huang G, Nammour R, Symes W. Full-waveform inversion via source-receiver extension. Geophysics. 2017;**82**: R153-R171

[37] Biswas R, Sen MK. 2D Full Waveform Inversion and Uncertainty Estimation using the Reversible Jump Hamiltonian Monte Carlo. SEG Technical Program Expanded Abstracts. 2017;**87**:1280-1285

[38] da Silva NV, Yao G, Warner M. Semiglobal viscoacoustic full-waveform inversion. Geophysics. 2019;**84**:R271-R293

[39] Zhang Z-d, Alkhalifah T. Local-crosscorrelation elastic full-waveform inversion. Geophysics. 2019;**84**:R897-R908

[40] Sen MK, Roy IG. Computation of differential seismograms and iteration adaptive regularization in prestack waveform inversion. Geophysics. 2003; **68**:2026-2039

[41] Padhi A, Mallick S, Fortin W, Holbrook WS, Blacic T. 2-D ocean temperature and salinity images from pre-stack seismic waveform inversion methods: an example from the South China Sea. Geophysical Journal International. 2015;**202**: 800-810

[42] Kennett BLN. Seismic Wave Propagation in Stratified Media. Cambridge, United Kingdom: Cambridge University Press; 1983

[43] Kennett BLN, Kerry NJ. Seismic waves in a stratified half space. Geophysical Journal of the Royal Astronomical Society. 1979;**57**:557-583

[44] Yilmaz O. Seismic data processing, Society of Exploration Geophysicists. Tulsa, OK, USA; 1987

[45] Castagna JP, Batzle ML, Eastwood RL. Relationships between compressional-wave and shear-wave velocities in clastic silicate rocks. Geophysics. 1985;**50**:571-581

[46] Hilterman F. Is AVO the seismic signature of lithology? A case study of Ship Shoal-South Addition. The Leading Edge. 1990;**10**:39-42

[47] Gardner GHF, Gardner LW, Gregory AR. Formation velocity and density – The diagnostic basics for stratigraphic traps. Geophysics. 1974;**39**: 770-780

[48] Todd CP, Backus MM. Offset-dependent reflectivity in a structural context. SEG Technical Program Expanded Abstracts. 1985;**55**:586-588

[49] Resnick JR. Seismic data processing for AVO and AVA analysis. In: Castagna JE, Backus MM, editors. Offset Dependent Reflectivity—Theory and Practice for AVO and AVA analysis. Tulsa, OK, USA: Society of Exploration Geophysicists; 1993. pp. 175-189

[50] Mukhopadhyay PK, Mallick S. An accurate ray-based offset-to-angle transform from normal moveout uncorrected multicomponent data in a transversely isotropic medium with vertical symmetry axis. Geophysics. 2011;**76**:C41-C51

[51] Shaw R, Srivastava S. Particle swarm optimization: A new tool to invert

geophysical data. Geophysics. 2007;**72**: F75-F83

[52] Tronicke J, Paasche H, Böniger U. Crosshole traveltime tomography using particle swarm optimization: A near-surface field example. Geophysics. 2016; **77**:R19-R32

[53] Mallick S. A simple approximation to the P-wave reflection coefficient and its implication in the inversion of amplitude variation with offset data. Geophysics. 1993;**58**:544-552

[54] Koefoed O. On the effect of Poisson's ratios of rock strata on the reflection coefficients of plane waves. Geophysical Prospecting. 1955;**3**:381-387

[55] Koefoed O. Reflection and transmission coefficients for plane longitudinal incident waves. Geophysical Prospecting. 1962;**10**:304-351

[56] Bortfeld R. Approximations to the reflection and transmission coefficients of plane longitudinal and transverse waves. Geophysical Prospecting. 1961;**9**:485-502

[57] Richards PG, Frasier CW. Scattering of elastic waves from depth-dependent inhomogeneities. Geophysics. 1976;**41**: 441-458

[58] Shuey RT. A simplification of the Zoeppritz, equations. Geophysics. 1985; **50**:609-614

[59] Aki K, Richards PG. Quantitative Seismology. Herndon, VA, USA: University Science Books; 2002

[60] Sen MK, Stoffa PL. Nonlinear one-dimensional seismic waveform inversion using simulated annealing. Geophysics. 1991;**56**:1624-1638

[61] Sen MK, Biswas R. Transdimensional seismic inversion using the reversible jump Hamiltonian Monte Carlo algorithm. Geophysics. 2017;**82**:R119-R134

[62] Zhu D, Gibson R. Seismic inversion and uncertainty quantification using transdimensional Markov chain Monte Carlo method. Geophysics. 2018;**83**: R321-R334

[63] Jian W, Lei Z, Hao C, Xiu-ming W. Trans-dimensional Bayesian inversion for directional resistivity logging while drilling data. SEG Technical Program Expanded Abstracts. 2018;**88**:849-852

[64] Visser G, Guo P, Saygin E. Bayesian transdimensional seismic full-waveform inversion with a dipping layer parameterization. Geophysics. 2019;**84**: R845-R858

[65] Dhara A, Sen MK, Yang D, Schmedes J, Routh P, Sain R. Facies and reservoir properties estimation by a transdimensional seismic inversion. SEG Technical Program Expanded Abstracts. 2020;**90**:255-259

[66] Deb K, Agrawal S. A niched-penalty approach for constraint handling in genetic algorithms. Vienna: Springer; 1999. DOI: 10.1007/978-3-7091-6384-9_40

[67] Deb K, Pratap A, Agarwal S, Meyarivan T. A fast and elitist multi-objective genetic algorithm: NSGA-II. IEEE Transactions on Evolutionary Computation. 2002;**6**(2):181-197

Section 2

Societal Applications

Chapter 2

Application of Genetic Algorithms in Health Sciences

Rohollah Fallah Madvari

Abstract

In this section, we introduce genetic algorithm (GA) and some of its applications in various health fields. Although GA and some other meta-heuristics are inspired by biology, they are more familiar to experts in other sciences, and these methods are often used to solve complex problems. The use of GAs has promising implications in various health specialities, including occupational health, environmental health, HSE, occupational medicine, industrial safety, ergonomics, toxicology, health care management, etc. This section of the book presents applications of GAs in disease screening, diagnosis, prognosis and health care management, and enables professionals to envision possible applications of this meta-heuristic method in their health professions. In the following, we discuss some applications of GAs in predicting, measuring and controlling factors that affect health.

Keywords: health sciences, genetic algorithms, optimisation, health, safety, environment

1. Introduction

The GA is an evolutionary method for constrained and unconstrained optimisation problems that uses the principles of Darwin's natural selection to find the optimal formula to predict or match the model. In general, GA are based on iteration, with most of the parts being selected as random processes. In nature, better generations result from the combination of better chromosomes. In the meantime, there are sometimes mutations in the chromosomes that can make the next generation better. GA also use this idea to solve problems. The GA starts its entire process with an initial population of random samples. Each sample in the population represents a potential solution to the problem at hand. The samples are evolved through successive iterations, called generations, and evaluated against fitness criteria during each generation. The population of the next generation is built by genetic operators and the iteration process continues until the final state is reached.

In the following, we discuss some applications of GA in predicting, measuring and controlling factors affecting health.

2. Application of genetic algorithm in acoustics

Noise is an unwanted and uncomfortable sound that has been the focus of much research as a harmful factor. Many studies have been conducted on the effects of sound, most of which point to the negative effects of sound on health [1, 2].

In general, the adverse effects of sound can be divided into three categories: psychological effects, interference with activities and physiological effects [1, 3]. Factors such as the level of sound exposure, the duration of exposure and the frequency spectrum determine the extent and nature of sound effects.

In recent years, metal foams have emerged as an attractive area of research from a scientific, industrial and audio application point of view [4]. Acoustic absorption is one of the most important functional properties of metal foams [5]. Porous metal is one of the most promising materials because porous metal has higher mechanical strength and hardness, resistance to heat, corrosion and weathering than non-metallic porous materials such as glass wool and urethane foam [6]. The sound absorption behaviour of porous metal depends on the cell structure, which is mainly divided into two types: open-cell structure and closed-cell structure [7]. Porous metals with excellent absorption properties have an open cell structure because the sound wave propagates inside the material. On the other hand, a porous metal with a closed cell, which has a wavy cell wall, does not absorb sound. Sound absorption occurs due to air adhesion friction at the boundary between the matrix and the air, and part of the sound energy is converted into thermal energy. Therefore, the absorption of sound by porous metal is related to the behaviour of air diffusion inside the cell, and therefore the characteristic of sound absorption by porous metal is strongly dependent on its cell structure, which is determined by the manufacturing methods and conditions [8, 9]. Many parameters can affect the sound absorption coefficient of aluminium foams, including porosity (Ω), pore size (D), pore opening (d), thickness (t), resistance to static flow, etc. (**Figure 1**) [10–12].

Various optimisation algorithms have been proposed, including the genetic algorithm (GA) [13]. The GA algorithm checks the neighbourhood by repeatedly expanding the search domain in the neighbourhood of the current solution and moving from the current solution to the increasing neighbourhood. This process is repeated until the current solution cannot be improved. This process continues until the optimal location is reached.

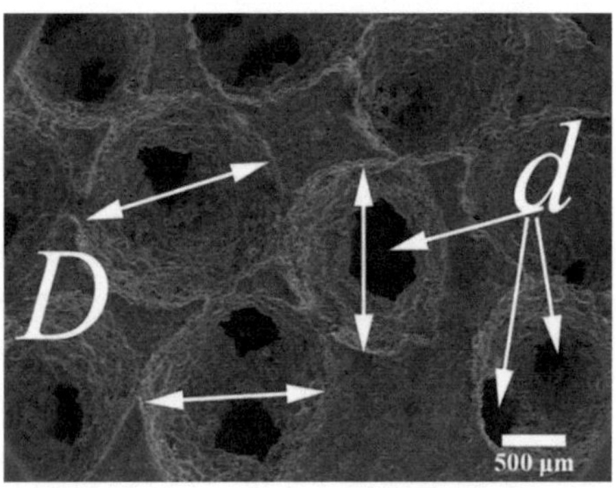

Figure 1.
The morphology of the metal foam sample, where D represents the pore diameter and d represents the opening diameter of the pores [12].

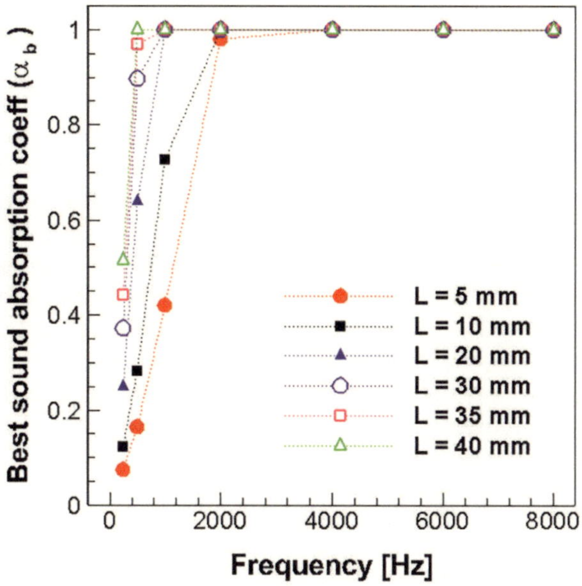

Figure 2.
Optimised sound absorption coefficient chart at various frequencies using GA in the thickness of 5–40 mm [13].

GA is a problem domain-independent method and quickly searches the search space for an optimised point with a quality function. GA has a distinct advantage over other stochastic methods. It is very easy to parallelise the algorithm. This is because the calculations of each iteration are independent of each other.

In general, the (Ω), (D), (d) and (t) are not regulated in the manufacture of sound-absorbing foam [14]. Therefore, approximate sizes of existing foam are tested through trial and error to find the best sound-absorbing foam. However, if the shape of the metal foam can be pre-determined for constant sound absorption with an optimum SAC, it is a great step towards the intelligent production of porous foam. We discuss how to use GA to improve an optimal set of metal foaming parameters including the (Ω), (D) and (d) at any thickness and frequency to obtain an optimal foam [13]. In **Figure 2**, the value of the SAC optimised for different thicknesses is shown for each frequency.

The results in **Figure 2** show that, in order to increase the amount of sound absorption, the thickness of the panel must be increased at frequencies below 2000 Hz. However, at higher frequencies, for each thickness, it is possible to find conditions (the value of d, D, Ω), so that the amount of sound absorption reaches the maximum.

3. Optimisation of sound power transaction of multi-wall panels using genetic algorithm

Tanyo et al. [15] used a GA to optimise the layout of structural layers according to the number of layers and corresponding thicknesses. The optimisation process was performed by selecting materials from a given list that included a limited number of solid, liquid and foam materials. One of the advantages of this approach is that there is no need to create new materials for a material designed for a specific application.

The research by Shojaeifard et al. [16] began by explaining the theory behind the acoustic analysis of multilayer panels using the transfer matrix method. After comparing with the existing laboratory results in this field and ensuring the correctness of the modelling, in the next step, using this valid model, a multi-objective optimisation of the acoustic behaviour of the panel was carried out. The acoustic behaviour of the multi-wall acoustic panel is optimised by finding the optimal material and thickness for the layers, and at the same time considering the acoustic behaviour criteria, minimising the weight of the panel is considered during the optimisation process. Since in the present work, the optimisation variables include both continuous and discrete variables, therefore the GA, as one of the most powerful algorithms that can manage continuous and discrete variables together, has been selected to solve the present problem.

4. Using genetic algorithm in multi-objective optimisation of external louvres in office buildings

Inadequate lighting will be associated with some degree of perceptual error such as sleepiness [17]. Improving the comfort and convenience of the interior space by optimising the level of natural light is one of the most important issues in the renovation and improvement of space, especially in office buildings [18]. With the increase in energy consumption, the need for multi-objective optimisation and efforts to reduce consumption is increasing, especially in developing countries, and this has led designers towards excellent architecture and maximum use of renewable energy, optimal use of energy and lighting. During the day, the sun not only provides a favourable environment for users but also reduces the energy used to cool and heat the environment. As a result, one of the most important factors in improving the energy efficiency of the building is to control the amount of light entering the space, and considering that the only part of the building that directly receives sunlight from the sun is the window, the use of louvres to control the amount of sunlight entering the space is essential. The use of daylight in many cities in Iran, including the city of Tehran, is remarkable due to its favourable geographical location and the availability of many sunny hours throughout the year [18]. The purpose of this research is to apply artificial intelligence and algorithmic programming to estimate the proportions and technical specifications of external louvres and to propose a model for the design of southern openings of office space in the direction of efficiency and intelligent consumption of energy and providing the required level of light in the interior space. The research method in this research is simulation and logical reasoning, therefore, using office space simulation, parametric design of louvres and optimisation of parameters [degree of rotation (θ), length (x), distance from the window (a), amount of reflection (R) and number of louvres (n)] using GA to design the south window has been analysed and studied in accordance with the conditions of solar radiation in Tehran. The results show that the use of external louvres is very efficient in controlling and improving the quality of light (**Figure 3**) [18].

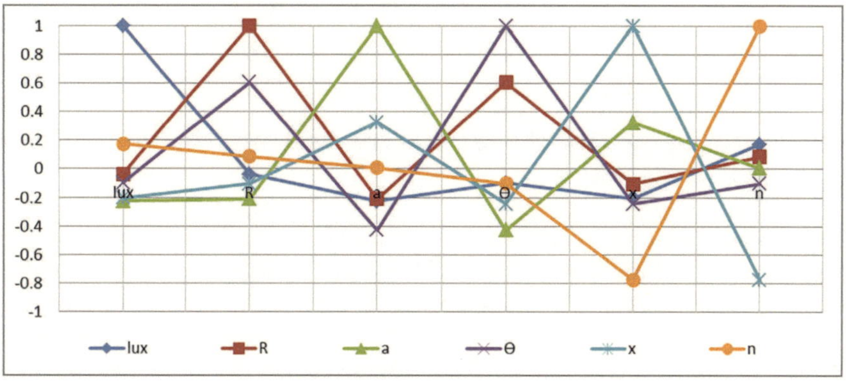

Figure 3.
The results of the optimised brightness of the simulated sample [18].

5. Multi-objective optimisation of window shape in order to simultaneously provide visual comfort and energy efficiency components through genetic algorithm

Zhang et al. [19] used a multi-objective GA to optimise the thermal and daylighting performance of a school building in the cold climate of Tianjin, China. First, different plan designs, i.e. one-way open corridors, one-way closed corridors and two-way corridors, were considered the main examples because they represent the structures of school buildings in China. The main results show that the best case is related to the design of the two-way corridor, mainly towards the south, in order to take advantage of the absorption of radiation during winter. The Galapagos algorithm and Octopus are optimisation tools in the Grasshopper environment [20]. Galapagos algorithm as an extension for the single-purpose GA execution engine and Octopus algorithm as an extension for the multi-objective GA execution engine in Grasshopper software. As a multi-objective optimisation tool, the GA helps the production of plans and forms to move towards the optimum, and even when the local optimum occurs, it will cause a deviation from it by creating a genetic mutation [21]. For single objective optimisation, the objective function is the minimum or maximum value of each of these performance indicators. The GA is then used to explore the relationship between the design components of the window shape and the daylight and energy performance indicators and to generate new design options for better performance. Multi-objective optimisation involves finding intermediate solutions for different objectives. Firoozeh and Sayyed Majid [21] multi-objective optimisation of window shape to simultaneously provide the components of visual comfort and energy efficiency through GA provide the components of visual comfort (increasing brightness and reducing glare) and energy efficiency (reducing energy consumption). In order to achieve the optimal values of window design parameters, the optimisation process based on simulation through GA was performed automatically in Grasshopper software as multi-objective. Finally, multi-objective optimisation by visualising the boundaries of the solution space can significantly reduce the complexity of the problem and help the designer to achieve a set of variables that simultaneously consider relatively good values of all the objective functions and the possibility of choosing options with the priority of each of the conflicting objectives to achieve a match between the project expectations and the final design [21].

6. Optimisation of the roof of a three-story residential building with the help of genetic algorithm

The walls of a building are important in terms of heat exchange and control of the building's energy consumption, as they are the outermost envelope of the building in direct contact with the air and temperature changes [22]. The body of the roof is more important than the other walls of the building because its heat exchange is more exposed to sunlight and other factors than other walls due to its area and time. The aim of this section is to reduce energy consumption in a residential building in Shiraz city and to achieve thermal comfort in the building spaces by optimising the roof of the building. In this context, the following question was raised What is the most optimal design of the roof of the building (materials and passive design methods) so that the energy consumption of the building is minimised and the spaces of the building are placed within the range of thermal comfort? The amount of reduction in the energy consumption of the building has also been considered in this research. The research method is quantitative and the energy consumption of the building and the thermal comfort index were done with Energy Plus software and the optimisation process was done with GA [22]. The building roof variables were extracted and defined in three general categories: passive energy system, physical characteristics of the roof and location of the roof. The results obtained from the building simulation calculations and the objective function output of the GA showed that the best roof models provided reduced energy consumption by 50% and the average thermal comfort index was 0.9 and 0.68, respectively in the warmest and coldest months of summer and winter.

The GA is one of the methods used in many architectural optimisation projects as an optimisation method when the number of variables is large [22]. Compared to other optimisation methods, this algorithm has advantages that have made it used in different fields, including the fact that it deals with discrete values and is not limited to continuous values. The GA is inspired by the idea of natural evolution and Darwin's principle of survival. In general, GAs are population-based algorithms that can be classified as global optimisation algorithms, which include the operation of searching for the global optimal solution as well as the operation of improving local solutions. The iterative process of GAs leads to better solutions based on mating and crossing of higher-performing parents. The genetic coding of humans is called the genotype, and the coded information of individual characteristics is called the phenotype. The operations related to the genetic coding of the parents produce results for the next generation. A basic GA has three main operations that are performed at each iteration: inheritance, crossover and mutation [22].

7. Accurate modelling and prediction of $PM_{2.5}$ concentration or use of genetic algorithm

Many large cities face the problem of air pollution, partly from mobile sources such as vehicles and partly from stationary sources such as industry [23]. The gradual and long-term effects of air pollution have led authorities and people to pay less attention to it. It should be noted that the increasing trend in the number of deaths, cancers and heart attacks caused by air pollution indicates that air pollution has caused the gradual death of people [23].

There are limitations in measuring air pollution concentration, one of the important limitations is the number and spatial distribution of air pollution monitoring stations. Due to the high cost and lack of necessary facilities, it is not possible to cover the entire city with air pollution monitoring stations; therefore, interpolation and prediction algorithms are used to model air pollution. The accuracy of these models is of particular importance and various research works have been conducted to improve the quality of air pollution modelling.

It is used to select effective parameters for estimating the concentration of air pollution. This algorithm uses biogeography-based optimisation and GA for optimisation [23].

8. Optimising air distribution in ventilation network by using genetic algorithm method

Ventilation is one of the most important support activities in the production chain of underground mining and construction activities, providing and distributing the air required by the various sectors [24]. The main purpose of a ventilation system is to economically supply the required fresh air at a speed and volume sufficient to rapidly dilute and remove contaminants from work areas. Pollutants and disturbance factors in drilling and mining operations include a wide range of flammable and noxious gases, dust, toxic gases from the explosion of explosives, heat and moisture and the acceptable levels of these pollutants are defined by a number of standards. Inadequate ventilation during the mine's operating life can cause problems with extraction and even halt production. The design of ventilation systems is usually based on technical considerations and economic analysis is rarely used as a decision criterion.

Optimum air distribution in mining operations can be described as the most important practical solution for reducing the operating and capital costs of the ventilation network, which can be achieved by selecting the correct position and resistance of the regulating doors, as well as the position and specifications of the booster fans, for real and adequate distribution of airflow in the network. It can be achieved.

In order to optimise the air distribution, a GA can be used to search for the optimum values of fan allocation, damper pressure drop and flow intensity of each branch of the ventilation network.

The GA was determined by analysing the optimal fan power and the amount of pressure drop required for the control dampers. The effect of the GA parameter values of mutation rate and grafting rate, together with the population size, on achieving the optimum response was then investigated. It was found that increasing the population to a certain extent increases the probability of achieving the optimal solution. As the linkage and mutation coefficients increase, the accuracy of the calculations decreases and the time to reach the optimal solution also increases. To complete the research, it is proposed to investigate GA coding in combination with the Hardy Cross method for the optimal design of air distribution in a mine ventilation network [24].

9. Scheduling employees' work shifts using a genetic algorithm approach

The general objective of this section is to apply human factors engineering to scheduling theory in order to exploit the optimal performance of employees [25]. The problem of scheduling the work shift of employees with variable performance is investigated in this section. The objective function of the mathematical model for employee scheduling presented in this section is to minimise labour costs and attempts to assign efficient employees to work shifts in order to meet the work demand of the organisation. The important feature of the presented mathematical model is the consideration of ergonomic dimensions of employees, including learning, forgetting and fatigue caused by work. GA was used to solve the presented mathematical model in a reasonable computation time. In order to verify the efficiency and effectiveness of the GA compared to the exact problem-solving methods, the performance of the algorithm was compared with the performance of the LINGO software and the lower limit of the presented examples. The results of the study by Akbari et al. showed that the presented model has the ability to model human factors and provides favourable work shifts. This study also showed that the human parameters studied have an impact on the efficiency of the employees and consequently on the planning of the organisation's work schedules. Therefore, it is suggested that managers in organisations should use the proposed model to study the effect of human factors on employees' efficiency and provide an optimal schedule for employees' work shifts [25].

The set of analyses showed that the change in human parameters has an effect on the planning of work shifts, and therefore it is suggested that managers use the presented model to study these effects and provide suitable work schedules for employees. The presented model also has the ability to be used in a dual way to plan job rotation with the aim of improving employees' health, which can be investigated in future research. In general, the results showed that the effect of human parameters on the efficiency of employees and organisations in scheduling work shifts can be mathematically modelled using exponential and hyperbolic functions, and the presented GA has the necessary power to solve the optimisation model. Modelling other human factors such as motivation, stress, etc. in employee scheduling problems and applying other meta-heuristic methods and comparing the results are topics that can be explored in future research.

10. Providing the optimal model for urban waste management system using genetic algorithm based on fuzzy logic

In recent years, all kinds of models have been studied and used to evaluate the waste management subsystems of the city of Tehran and to select the best waste management option [26]. However, the problem of final waste disposal in Tehran is one of the most important issues related to the environmental management of this metropolis. The aim of the present research is to provide a model to allocate the optimal annual amount of Tehran's waste to waste management subsystems in order to achieve maximum efficiency, reduce costs and increase the system's revenue. Firstly, the data required for the research was collected by referring to the Aradkoh Complex in Tehran and personal interviews with experts and using the information recorded in the Aradkoh Complex. Then, the proposed research model for the purpose of allocating the optimal amount of annual waste taking into account all constraints to five recycling subsystems, aerobic compost, anaerobic digester, waste incinerator and sanitary landfill using the GA based on fuzzy logic with the aim of reducing the total cost of waste management system Shahri was implemented in MATLAB environment and its results were analysed. According to the results of the optimal model proposed by the research, it is necessary to allocate the optimal flow and process of the annual waste of Tehran city among recycling, aerobic composting, anaerobic digestion, waste incineration and sanitary landfill systems with more accuracy in order to increase the annual efficiency of the waste management system. The city of Tehran should be followed.

In this research, it was tried to present an optimal model for Tehran's urban waste management system with the aim of reducing costs and increasing income using the improved GA method by fuzzy logic controller, and based on this, the annual optimal values of Tehran's waste were allocated to each of five under the recycling system, aerobic compost, anaerobic digester, waste incinerator and sanitary landfill, so that the presented model can be fully implemented to increase the annual efficiency of Tehran's waste management system. The results of this research showed that the optimal model of the urban waste management system is a combination of different waste management options in order to achieve maximum productivity and reduce costs and increase revenues, and using only one waste management option is not cost-effective. It can also be said that increasing the capacity allocated to each of the subsystems does not mean reducing costs and increasing revenue generation, and the most optimal point in each of the subsystems is the point where the objective function is optimised [26].

According to the results of the implementation of the proposed research model using fuzzy GA, increasing the capacity of subsystems with lower costs and higher revenues up to a certain amount can lead to the highest profitability and the capacity increase of these subsystems after this amount is not specified. It can be affordable. Also, by examining the changes in the total cost of the system, it can be concluded that increasing the capacity of the subsystems in total up to a certain amount will optimise the urban waste management system, and after that, it will not have a positive effect on reducing the cost of the system and increasing income generation. The comparison of the results of the GA implementation alone with the GA improved by the fuzzy logic controller shows that the fuzzy system plays a positive role in the efficiency of the GA in reaching a more optimal solution with a higher speed, and therefore it can be used to make the algorithm more dynamic. Genetics makes good use of the fuzzy system in this model in the direction of greater efficiency to reach the optimal solution.

11. Conclusion

The idea of a GA, like an artificial neural network, is inspired by nature. Another such innovation is evolution. By "simulating" the process of evolution in nature, GAs search the "space of candidate solutions" to find the best possible solution to a problem. In the search for the optimal solution, a set or population of initial solutions is first generated. Then, in successive "generations", a set of modified solutions is produced (in each generation of the GA, certain changes are made to the genes of the chromosomes that make up the population). The initial solutions are usually modified in such a way that, in each generation, the population of solutions "converges" towards the optimal solution.

This branch is inspired by the field of "artificial intelligence", which is based on the mechanism of evolution of living organisms and the production of more successful and graceful species in nature. In other words, the main idea of GAs is "survival of the fittest". In general, it is an algorithm based on repetition, most of its parts are selected as random processes, and these algorithms consist of the functional parts of adaptation, representation, selection and modification. GAs can help many health sciences and even medicine in the future.

Author details

Rohollah Fallah Madvari
Department of Occupational Health Engineering, School of Public Health, Shahid Sadoughi University of Medical Sciences, Yazd, Iran

*Address all correspondence to: fallah134@gmail.com

References

[1] Fallah Madvari R, Farhang Dehghan S, Abbsi M, Laal F, Fallah Madvari AR, Haji Moradi F, et al. The Relationship between sound pressure level with cognitive failure indicators and noise injury in a ceramic industry. Iran Occupational Health Journal. 2020;**17**(1):460-474

[2] Fallah Madvari R, Zare Sakhvidi MJ, Jafari Nodoushan M, Askari J, Fallahzadeh H, Raiszade DM. Effect of sound pressure levels on problem-solving abilities with the mediation of personality traits. Hearing, Balance and Communication. 2023;**21**(3):194-200

[3] Fallah Madvari R, Tahmasbi Abdar F, Halvani GH, Sefidkar R, Mohammadi M, Sojoudi S, et al. Correlation between noise exposure and mental health components among Iranian Steel Workers, 2021. Journal of Occupational Health and Epidemiology. 2022;**11**(2):148-156

[4] Azizan MA, Ismail MH, Salleh NAM, Natarajan DV. Sound absorption properties at high sound frequency of open cell aluminium foam. Journal of Mechanical Engineering. 2017;**1**:161-173

[5] Han F, Seiffert G, Zhao Y, Gibbs B. Acoustic absorption behaviour of an open-celled aluminium foam. Journal of Physics D: Applied Physics. 2003;**36**(3):294

[6] Hakamada M, Kuromura T, Chen Y, Kusuda H, Mabuchi M. Sound absorption characteristics of porous aluminum fabricated by spacer method. Journal of Applied Physics. 2006;**100**(11):114908

[7] Gibson L, Ashby M. Cellular Solids: Structure and Properties (2nd ed., Cambridge Solid State Science Series). Cambridge: Cambridge University Press; 1997. DOI: 10.1017/CBO9781139878326

[8] Madvari RF, Jafari MJ, Hong TW, Laal F, Sharak MN. Effect of porosity, pore size, and pore-opening size optimized on the sound absorption coefficient of aluminum foam. Noise Control Engineering Journal. 2023;**71**(2):92-100

[9] Jafari MJ, Madvari RF, Ebadzadeh T. Optimized design and experimental validation of sound absorption coefficient performance in aluminium metal foam by spark plasma sintering. Heliyon. 2023;**9**(6):e16428

[10] Guan D, Wu JH, Wu J, Li J, Zhao W. Acoustic performance of aluminum foams with semiopen cells. Applied Acoustics. 2015;**87**:103-108

[11] Jafari MJ, Khavanin A, Ebadzadeh T, Fazlali M, Sharak MN, Madvari RF. Optimization of the morphological parameters of a metal foam for the highest sound absorption coefficient using local search algorithm. Archives of Acoustics. 2020;**45**(3):487-497

[12] Li Y, Wang X, Wang X, Ren Y, Han F, Wen C. Sound absorption characteristics of aluminum foam with spherical cells. Journal of Applied Physics. 2011;**110**(11):113525

[13] Jafari MJ, Sharak MN, Khavanin A, Ebadzadeh T, Fazlali M, Madvari RF. Improving the cellular characteristics of aluminum foam for maximum sound absorption coefficient using genetic algorithm. Sound & Vibration. 2021;**55**(2):117-130

[14] Madvari RF, Sharak MN, Tehrani MJ, Abbasi M. Estimation of metal foam

microstructure parameters for maximum sound absorption coefficient in specified frequency band using particle swarm optimisation. Archives of Acoustics. 2022;**47**:33-42

[15] Tanneau O, Casimir J, Lamary P. Optimization of multilayered panels with poroelastic components for an acoustical transmission objective. The Journal of the Acoustical Society of America. 2006;**120**(3):1227-1238

[16] Shojaeifard M, Talebitooti R, Torabi M, Ahmadi R. Optimization of power transmission interaction of multilayered panel using genetic algorithm. Modares Mechanical Engineering. 2014;**14**(1):27-34

[17] Madvari RF, Sefidkar R, Halvani GH, Alizadeh HM. Quantitative indicators of street lighting with mood, fatigue, mental workload and sleepiness in car drivers: Using generalized structural equation modeling. Heliyon. 2023;**9**(1):e12904

[18] Mahdavinejad M, arbab m, arbab m. Genetic algorithm for multi-objective optimization external louvers in high-performance office buildings. Journal of Architectural Thought. 2019;**3**(5):214-235

[19] Zhang A, Bokel R, van den Dobbelsteen A, Sun Y, Huang Q, Zhang Q. Optimization of thermal and daylight performance of school buildings based on a multi-objective genetic algorithm in the cold climate of China. Energy and Buildings. 2017;**139**:371-384

[20] Roudsari MS, Pak M. Smith A, et al. Ladybug: A parametric environmental plugin for grasshopper to help designers create an environmentally-conscious design. In: Proceedings of the 13th International IBPSA Conference held in Lyon, France; 2013

[21] Ma X, Firoozeh O, Majid MSS. Multi-objective optimization of window configuration to provide integrated visual comfort components and energy efficiency by the Genetic algorithm (the case study: primary school classroom in Tehran-Iran). Geographical Journal of Territory. 2021;**17**(68):1-20. Available from: https://sid.ir/paper/951351/en

[22] Keshtkaran P, Movahed K, Barzegar Z. Roof optimization of three floor residential building using G.A case study: Shiraz, Iran. Journal of Sustainable Architecture and Urban Design. 2022;**10**(1):141-167

[23] Gholami A, Ebrahimian GY. Spatiotemporal estimation of PM2.5 concentration using remotely sensed data, machine learning, and optimization algorithms. Journal of Geomatics Science and Technology. 2023;**12**(2):136-151

[24] Shahabi RS, Larijani H, Zeyni EE, Sadeghzadeh MH. Optimization of air distribution in mine ventilation networks based on genetic algorithm (case study: Kalariz Coal Mine). Tunneling & Underground Space Engineering. 2019;**8**(2):143-166

[25] Akbari M, Dorri Nokarani B, Zandieh M. Scheduling working shifts for multi-skilled workforces with genetic algorithm approach. Journal of Industrial Management Perspective. 2012;**2**(3, Autumn):87-102

[26] Ahani M, Arjmandi R, Hoveidi H, Ghoddousi J, Miri Lavasani MR. Providing optimal model for municipal solid waste management system using genetic algorithm based on fuzzy logic (case study: Tehran City). Journal of Environmental Science and Technology. 2021;**23**(1):27-40

Chapter 3

Using Group Theory to Generate Initial Population for a Genetic Algorithm for Solving Traveling Salesman

Dharm Raj Singh

Abstract

In this chapter, we propose a novel algorithm that uses Genetic algorithm with group theory for initial population generation and also propose a novel crossover for solving Traveling Salesman Problem. In the group tour construction method, each individual/initial tour has distinct start city provided that population size is equal to total number of cities. In the initial population, each individual/tour has a distinct starting city. The distinct starting cites of each tour provide genetic material for exploration for the whole search space. Therefore, a heterogeneous starting city of a tour in initial population is generated to have rich diversity. Proposed crossover based on greedy method of sub-tour connection drives the efficient local search, followed by 2-opt mutation for improvement of tour for enhanced/optimal solution. The result of the proposed algorithm is compared with other standard algorithms followed by conclusion.

Keywords: genetic algorithms, traveling salesman problem, group theory for population generation, 2-opt mutation, group theory

1. Introduction

1.1 Genetic algorithm (GA)

Genetic algorithm draws the idea from natural selection and natural genetics principles for searching and optimization algorithm. In this method, we use survival of the fittest rule of natural evolution. Invention of GA was done in 1960 by John Holland [1]. It consists of population of chromosomes, with each chromosome representing a solution to the particular problem. Each chromosome is evaluated to obtain its fitness value of the chromosome against some given fitness function. A set of chromosomes are selected for genetic operation(s) (selection, crossover, and mutation) in order to get new chromosomes. Chromosomes are selected according to their fitness values to reproduce/generate the next generation by genetic operations, which generate new chromosomes. To achieve this, two transformations, namely crossover (generates new chromosomes by overlapping genes of two chromosomes) and mutation (creates a new chromosome by making changes of genes in a single chromosome), are used. After performance of crossover and mutation operation, we generate a new chromosome called child. The process continues by selecting fit chromosomes from parent and child population. Whole process of genetic algorithm is repeated until best individual is obtained or desired number of iterations completed, providing an optimal/suboptimal solution to the problem [2].

We developed a genetic algorithm for Traveling Salesman Problem (TSP) to provide balance between exploration and exploitation for the search space. For this, all the components of the genetic algorithms were carefully examined.

1.2 Literature review

Traveling Salesman Problem is a famous combinatorial optimization problem, which is still NP-complete [3, 4]. There is no clear evidence for its origin. However, credit goes to Irish mathematician Hamilton and British mathematician Thomas Krikman for its mathematical formulation, which is discussed in detail in "Graph Theory 1736-1936" book by Wilson, Biggs and Lloyd 2 [5]. The general form was firstly studied by mathematician Karl Menger [6]. Alexander Schrijver [7] pointed out the connection between the works of Whitney and Menger along with growth of TSP in his paper "On the history of combinatorial optimization (till 1960)." Merrill Flood [8] while searching for the solution of school bus routing problem used TSP mathematically for the first time. Hassler Whitney [9] of Princeton University introduced the name traveling salesman problem. The popularity of TSP increased considerably during 1950s and 1960s when prizes were offered by RAND Corporation for solving steps of the problem. As a result of it, prime contributions were made by Fulkerson, Dantzig and Johnson [10] from the RAND Corporation who used branch and bound algorithm [11] for solving integer linear program for which they developed the cutting plane method [12] for its solution. Later 532 and 2392 city TSP was solved by M. Padberg and Rinaldi [13] in 1987, and 1000 city TSP by solved M. Grotschel and O. Holland [14]. In 1991, Reinelt [15] introduced TSPLIB, which is still providing problem instances for TSP. In last few decades, many heuristic and meta-heuristic algorithms have been developed with some of the following notable algorithms: Ant colony optimization [16–25], Neural network [26–29], Self-organizing maps [30], Particle swarm optimization techniques [31, 32], Simulated annealing [33], Weed optimization [34, 35], Genetic algorithm [36–39].

The rest of chapter is divided into following sections: Details of Genetic algorithm are given in Section 1. Section 2 describes our proposed hybrid methods. In Section 3, experimental result is presented followed by conclusion in Section 4.

2. Proposed hybrid method

The methods used here are hybrid because we have used a proposed group theory tour construction algorithm and proposed crossover with 2-opt mutation Croes (1958) [40]. The framework of the proposed algorithm is shown in Algorithm 1.

The main idea of the first step is to generate a population of chromosomes (tours) by using proposed group theory approach. Clearly, each chromosome of the population is same but which start city unique providing rich diversity of genetic materials for exploration.

Fitness value of each chromosome in the population is calculated in the second step. In the third step, select two parent chromosomes (selected randomly) from the population and replace the first chromosome with minimum fitness value (tour cost). After that, apply proposed crossover operator on the selected two chromosomes with crossover probability rate. And finally, apply 2-opt mutation operator on selected parent chromosome or new pair of chromosomes generated after crossover, with mutation probability rate. Mutation operator helps in generate new population, which is then replacing new population with the previous population by the new population. Whole process is repeated until termination condition is satisfied.

Algorithm 1: Proposed Algorithm

Generate initial population of the tour with population size P using Group theory.
Gen = 1;
while (Gen $\leq N_{Gen}$) do.
Calculate the fitness of each tour in P.
Bs = Best tour in P;
Randomly select two parents S_1 and S_2 tour in P;
$S_{1new} = B_s$;
$S_{2new} = S_2$;
rnd1 = rand (0,1];
if (rnd1 < crossover probability rate (*pc*)) **then.**
Perform proposed crossover on selected two parents B_s and S_2 to generate two new children C_1 and C_2;
end if.
$S_{1new} = C_1$;
$S_{2new} = C_2$;
rnd2 = rand (0,1];
if (rnd2 < mutation probability rate (*pm*)) **then.**
Perform 2-opt optimal mutation operator on S_{1new} and S_{2new}
update new population P′;
P ← P′;
end if.
Gen = Gen + 1;
end while

2.1 Proposed group theory for population generation

There are various possible methods for generating the initial population [41–43]. One of the simplest ways is generating the initial population randomly using random number generator. Zhang [42, 43] proposed greedy tour construction heuristic with Karp-patching for feasible tour construction and used for solving Assignment problem [44]. We proposed the group tour construction heuristic for initial population generation. In this method, nodes of graph are label using group of integers, Z_n with integer modulo n operation

$$a +_n b = (a + b) \bmod n \tag{1}$$

3	4	5	6	7	8	9	10	1	2
4	5	6	7	8	9	10	1	2	3
5	6	7	8	9	10	1	2	3	4
6	7	8	9	10	1	2	3	4	5
7	8	9	10	1	2	3	4	5	6
8	9	10	1	2	3	4	5	6	7
9	10	1	2	3	4	5	6	7	8
10	1	2	3	4	5	6	7	8	9
1	2	3	4	5	6	7	8	9	10
2	3	4	5	6	7	8	9	10	1

Figure 1.
Population generated using Group Modulo.

where "$+_n$" is operator that represents addition modulo of n, and (a + b) represents the normal addition of integers. This helped in generating the group table shown in **Figure 1**. In group table, no two row or column elements in the same position are identical. The function used for generating initial population of chromosomes (P) is as follows:

$$P(a) = \mathrm{mod}((a + b), n) + 1. \quad (2)$$

where a represents population size whose value is from a = 1 to population size, and b = 1 to n (As mentioned in earlier, we are taking population size equal to number of cities, therefore population size = n). For n = 10, the initial population generated using Group theory is as follows: Each chromosome in the initial population being unique (group theory technique) provides a wide diversity of genetic materials for exploring of search space.

2.2 Proposed crossover operator for GA

Sharing information between a pair of chromosomes is called crossover [2]. In this process genes of parent's chromosomes are swapped to generate offspring. The selection of the parent chromosomes is with the possibility that good chromosomes may generate better offspring. Goldberg described several order-based operators, such as the Partially Matched Crossover (PMX) [45]. The order crossover (OX) was suggested by Syswerda [46]. The position-based crossover (PBX) was introduced by [39]. The cycle crossover (CX) was suggested by Oliver, Smith & Holland [47]. Freisleben & Merz introduced a distance preserving crossover (DPX) [37]. Inspired by DPX crossover, we propose a new crossover in this chapter. In the proposed crossover the cities that are identical for the same position in both parents (s_1 and s_2) will not change in child c_1 and c_2 as shown in **Figure 2**. The remaining cities will change accordingly Algorithm 2.

Algorithm 2: Algorithm for Proposed Crossover

```
c1 = zeros (1, n);
c2 = zeros (1, n);
for i = 1: n do
    for j = 1: n do
        if (s2(i)==s1(j)) then;
            c1(i) = s2(j);
        end if
      if (s1(i) == s2(j)) then
          c2(i) = s1(j);
      end if
    end for
end for
```

2.3 Example

Given a complete weighted graph with 5 nodes obtain weighted (cost) matrix of graph is in **Figure 3**.

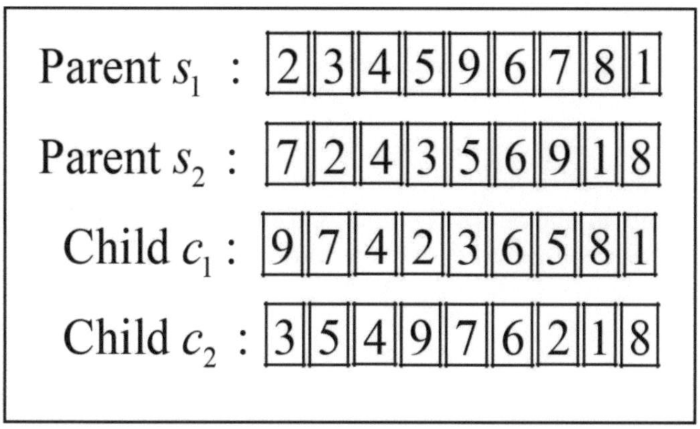

Figure 2.
Example of Proposed crossover.

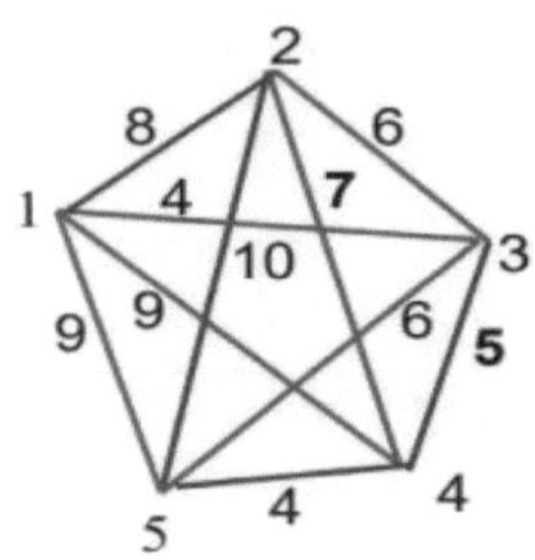

$$\begin{bmatrix} 0 & 8 & 4 & 9 & 9 \\ 8 & 0 & 6 & 7 & 10 \\ 4 & 6 & 0 & 5 & 6 \\ 9 & 7 & 5 & 0 & 4 \\ 9 & 10 & 6 & 4 & 0 \end{bmatrix}$$

Figure 3.
Example of a complete weighted graph with 5 nodes.

Population	cost
3 4 5 1 2	32
4 5 1 2 3	32
5 1 2 3 4	32
1 2 3 4 5	32
2 3 4 5 1	32

Figure 4.
Generate initial population with Population size =5 using Group Theory method.

Initial set Population size = 5, Crossover probability rate (pc) = 0.8 and Mutation probability rate (pm) = 0.2. Generate initial population with Population size = 5 using Proposed Group Theory method as shown in **Figure 4**.

Generate two random number between (1-Population size) is 4 and 2, then select two chromosome 4 and 2 from population after then replace first chromosome with minimum cost therefore selected chromosome with cost is

population					cost
2	3	4	5	1	32
3	4	5	1	2	32

Generate a random number rnd1 = 0.3243. if (rnd1 < = pc), then apply proposed crossover, after crossover generate two new chromosomes is

4 5 1 2 3
1 2 3 4 5

again, generate a random number rnd2 = 0.0161. if (rnd2 < = pm), then apply 2-opt mutation, the operation 2-opt mutation on both chromosome one by one given as Apply 2-opt mutation on first chromosome = 4 5 1 2 3

$$a = 3, b = 4, c = 5, d = 1, \text{and set } zmin = 0, i = 1, j = 1$$

$$z = dmat(a, c) + dmat(b, d) – dmat(c, d) – dmat(a, b);$$

Where dmat(a, c) represent cost from node a to c.

$$z = 6 + 9 – 9 - 5 = 1$$

if z < zmin false, then a = 3, b = 4, c = 1, d = 2, and i = 1, j = 4.

$$z = 4 + 7 – 8 - 5 = -2$$

if z < zmin true, then set zmin = −2, imin = 1, jmin = 4, and a = 3, b = 4, c = 2, d = 3, and i = 1, j = 5.

$$z = 6 + 5–6 - 5 = 0$$

if z < zmin false, then a = 4, b = 5, c = 1, d = 2, and i = 2, j = 4.

$$z = 9 + 10–8 - 4 = 7$$

if z < zmin false, then a = 4, b = 5, c = 2, d = 3, and i = 2, j = 5.

$$z = 7 + 6–6 - 4 = 3$$

if z < zmin false, then a = 5, b = 1, c = 2, d = 3, and i = 3, j = 5.

$$z = 10 + 4–6 - 9 = -1$$

if z < zmin false.

if zmin <0 Then apply 2-opt mutation between imin to jmin −1 on first selected chromosome = 4 5 1 2 3, after then we get a new chromosome is 1 5 4 2 3.

Again apply 2-opt mutation on new chromosome 1 5 4 2 3.

a = 3, b = 1, c = 5, d = 4, and set zmin = 0, i = 1, j = 3.

$$z = 6 + 9–4 - 4 = 7$$
$$z = \text{dmat}(a, c) + \text{dmat}(b, d)–\text{dmat}(c, d)–\text{dmat}(a, b);$$

if z < zmin false, then a = 3, b = 1, c = 4, d = 2, and i = 1, j = 4.

$$z = 5 + 8–7 - 4 = 2$$

if z < zmin false, then a = 3, b = 1, c = 2, d = 3, and i = 1, j = 5.

$$z = 6 + 4–6 - 4 = 0$$

if z < zmin false, then a = 1, b = 5, c = 4, d = 2, and i = 2, j = 4.

$$z = 9 + 10–7 - 9 = 3$$

if z < zmin false, then a = 1, b = 5, c = 2, d = 3and i = 2, j = 5.

$$z = 8 + 6–6 - 9 = -1$$

if z < zmin true, then set zmin = −1, imin = 2, jmin = 5, and a = 5, b = 4, c = 2, d = 3, and i = 3, j = 5.

$$z = 10 + 5–6 - 4 = 5$$

if z < zmin false.

if zmin <0, then apply 2-opt mutation between imin to jmin −1 on first selected chromosome = 1 5 4 2 3, after then we get a new chromosome is 1 2 4 5 3.

Again apply 2-opt mutation 1 2 4 5 3 and a = 3, b = 1, c = 2, d = 4, and set zmin = 0, i = 1, j = 3.

$$z = 6 + 9–7 - 4 = 4$$

if z < zmin false, then a = 3, b = 1, c = 4, d = 5, and i = 1, j = 4.

$$z = 5 + 9–4 - 4 = 6$$

if z < zmin false, then a = 3, b = 1, c = 5, d = 3, and i = 1, j = 5.

$$z = 6 + 4\text{–}6 - 4 = 0$$

if z < zmin false, then a = 1, b = 2, c = 4, d = 5, and i = 2, j = 4.

$$z = 9 + 10\text{–}4 - 8 = 8$$

if z < zmin false, then a = 1, b = 2, c = 5, d = 3 and i = 2, j = 5.

$$z = 9 + 6\text{–}6 - 8 = 1$$

if z < zmin false, then a = 2, b = 4, c = 5, d = 3, and i = 3, j = 5.

$$z = 10 + 5\text{–}6 - 7 = 2$$

if z < zmin false and completed loop, then we get new first child chromosome is 1 2 4 5 3.

Similarly Apply 2-opt mutation on second chromosome = 1 2 3 4 5 and completed loop, then we get new second child chromosome is 3 1 2 4 5.

After one iteration is completed, the new population is updated as

population					cost
1	2	4	5	3	29
3	1	2	4	5	29
5	1	2	3	4	32
1	2	3	4	5	32
2	3	4	5	1	32

3. Experimental results

3.1 Experimental setup

For evaluation purpose, results were generated on 2.20 GHz Intel Core i5 machine with 4 GB RAM.

3.2 Experimental design

The following standard benchmark data set was taken from TSPLIB: Eil51, berlin52, St70, Eil76, Pr76, Kroa100, Eil101, ch150, and ts225, were taken for performance comparison.

For the experiment, the experimental parameters crossover probability rate (*pc* = 0.8) and mutation probability rate (*pm* = 0.2) values were set to and respectively with number of iterations = 500. Population size of each instance equal to number of cities was generated by group tour construction method. For result comparison, Percentage Best Error (% Best Err.) is used. The Percentage Best Error is calculated as follows:

$$\text{PDbest} = \frac{[(\text{Best path cost from n trail}) - (\text{Best Known Solution(BKS)})]}{(\text{Best Known Solution(BKS)})} \times 100$$

3.3 Experimental result

For each standard TSP data taken, we performed n = 20 trails for first set of comparison shown in **Table 1** and **Figure 5**. Best results are shown in bold. Results are taken from [48]. It can be clearly seen that proposed method is better than Hierarchic method used for comparison for every dataset taken in to consideration. Proposed method although not exact but do provide good heuristic solution.

The pictorial presentation for performance comparison in terms of Mean (average) solution for different methods is shown in **Figure 5**. **Table 2** reports the outcome of some large size of instances. If size of instances is increases then increases the population size because population size of each instances is equal to number of cities in

	Proposed method			Hierarchic approach			BKS
Problem	**Best**	**Mean**	**% best error**	**Best**	**Mean**	**% best error**	
Eil51	**428.87**	**431.28**	**0.561**	431.74	443.39	3.39	428.87
Berlin52	**7544.37**	**7544.37**	**0.000**	7544.37	7544.37	0.000	7544.37
St70	**677.12**	**679.30**	**0.324**	687.24	700.58	3.47	677.11
Eil76	**544.37**	**553.05**	**1.184**	551.07	557.98	2.31	545.39
Pr76	**108160.00**	**108232.00**	**0.067**	113,798.56	115,072.29	6.39	108,159.44
Kroa100	**21285.00**	**21359.05**	**0.346**	22,122.75	22,435.31	5.40	21,285.44
Eil101	**645.25**	**656.62**	**4.391**	672.71	683.39	6.39	642.31
Ch150	**6588.60**	**6665.29**	**2.103**	6641.69	6677.12	2.21	6532.28
Tsp225	**3878.80**	**3909.04**	**1.297**	4090.54	4157.85	7.74	3859.00

Table 1.
Performance comparison of proposed algorithm and hierarchic algorithm.

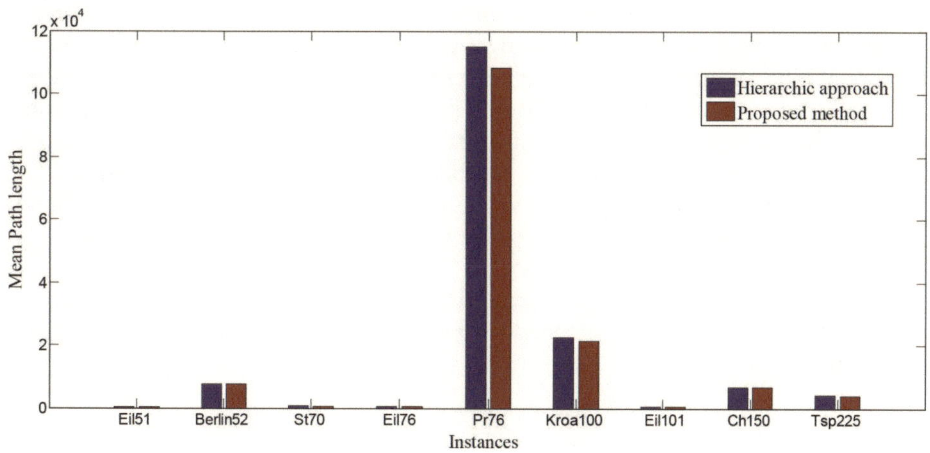

Figure 5.
Performance comparison of Average distance (Mean) (over 20 trails) between Proposed algorithm and Hierarchic algorithm for the 9 TSPLIB instances.

		Proposed method			BKS
Problem	**Best**	**Mean**	**% best error**	**time**	
Pr264	49,135	49388.90	0.00	37.98	49,135
a280	2594	2616.40	0.58	21.86	2579
Pr299	48,414	48732.05	0.46	70.59	48,191
lin318	42,446	42697.20	0.99	78.41	42,029
Pr439	108,767	109284.05	1.45	173.98	107,217
rat575	7128	7245.10	5.24	432.33	6773
Rat783	9302	9396.70	5.63	2934.03	8806
pr1002	267,663	270128.34	3.33	1698.14	259,047

Table 2.
Performance of proposed algorithm for large instances size.

instances therefore increases the time complexity and space complexity due to increases population size.

4. Conclusion

This chapter proposed a Genetic algorithm, which works by taking features of Group theory for tour construction, proposed crossover, and 2-opt mutation. Although the other reported methods can usually find a better solution for the TSP, their solution is dependent on the quality of the random initialization of the population and parameters. In order to have our proposed method uses heterogeneous population for process initialization with population size equal to number of cities, we applied group tour construction method. In group tour construction method generates all solution(tour) in the initial population has a distinct starting node that provides the initial exploration of the search space. After this, the proposed crossover and 2-opt mutation are applied. In order to maintain local optimality, crossover and mutation operators are used. By using crossover operator, new starting points were defined for a local search using information of the current population. The proposed crossover utilizes a greedy method for the duplicated paths in the parents for connecting subtours into the solution. However, 2-opt mutation can easily get stuck in a local optimum to improve the tour quality. The combination of the proposed method is required as 2-opt mutation easily gets stuck in local optimum. From the experimental results, one can easily find that our proposed algorithm gives better performance in comparison of [48].

As future work, we would like to extend the group theory for population initialization in other heuristics such as scheduling algorithm, network problems, estimation of distribution algorithm, etc.

Conflict of interest

The authors declare that they have no conflict of interest.

Author details

Dharm Raj Singh
Department of Computer Applications, Jagatpur P.G. College, Varanasi,
Uttar Pradesh, India

*Address all correspondence to: dharmrajsingh67@yahoo.com

References

[1] Holland JH. Adaptation in Natural and Artificial Systems. Ann Arbor: University of Michigan Press; 1975

[2] Sivanandam SN, Deepa SN. Genetic algorithms. In: Introduction to Genetic Algorithms. Berlin, Heidelberg: Springer; 2008. pp. 15-37

[3] Nemhauser GL, Savelsbergh MWP, Sigismondi GC. Constraint Classification for Mixed Integer Programming Formulations. Department of Mathematics and Computing Science, University of Technology; 1991

[4] Christos H. Papadimitriou and Kenneth Steiglitz, Combinatorial Optimization: Algorithms and Complexity. 1982

[5] Biggs NL, Lloyd EK, Wilson RJ. Graph Theory. Oxford University Press; 1986:1736–1936

[6] Menger K. Eine neue definition der bogenlange. Ergebnisse eines Mathematischen Kolloquiums. 1932;**2**:11-12

[7] Schrijver A. On the history of combinatorial optimization (till 1960). Handbooks in Operations Research and Management Science. 2005;**12**:1-68

[8] Flood MM. The traveling-salesman problem. Operations Research. 1956; **4**(1):61-75

[9] Whitney H. The mathematics of physical quantities: Part i: Mathematical models for measurement. The American Mathematical Monthly. 1968;**75**(2):115-138

[10] Dantzig G, Fulkerson R, Johnson S. Solution of a large-scale traveling-salesman problem. Journal of the Operations Research Society of America. 1954;**2**(4):393-410

[11] Laporte G. The traveling salesman problem: An overview of exact and approximate algorithms. European Journal of Operational Research. 1992; **59**(2):231-247

[12] Reinelt G. The Traveling Salesman: Computational Solutions for TSP Applications. Springer-Verlag; 1994

[13] Padberg M, Rinaldi G. Optimization of a 532-city symmetric traveling salesman problem by branch and cut. Operations Research Letters. 1987;**6**(1): 1-7

[14] Grotschel M, Holland O. Solution of large-scale symmetric travelling salesman problems. Mathematical Programming. 1991;**51**(1–3):141-202

[15] Reinelt G. TSPLIB—A traveling salesman problem library. ORSA Journal on Computing. 1991;**3**(4):376-384

[16] Chen S-M, Chien C-Y. Solving the traveling salesman problem based on the genetic simulated annealing ant colony system with particle swarm optimization techniques. Expert Systems with Applications. 2011;**38**(12):14439-14450

[17] Cecilia JM, et al. Enhancing data parallelism for ant colony optimization on GPUs. Journal of Parallel and Distributed Computing. 2013;**73**(1):42-51

[18] Deng W, Chen R, He B, Liu Y, Yin L, Guo J. A novel two-stage hybrid swarm intelligence optimization algorithm and application. Soft Computing. 2012; **16**(10):1707-1722

[19] Jun-man K, Yi Z. Application of an improved ant colony optimization on generalized traveling salesman problem. Energy Procedia. 2012;**17**:319-325

[20] Mahi M, Baykan OK, Kodaz H. A new hybrid method based on particle swarm optimization, ant colony optimization and 3-opt algorithms for traveling salesman problem. Applied Soft Computing. 2015;**30**:484-490

[21] Mavrovouniotis M, Yang S. Ant colony optimization with immigrants' schemes for the dynamic travelling salesman problem with traffic factors. Applied Soft Computing. 2013;**13**(10):4023-4037

[22] Rodrigo Pasti, Nunes de Castro L. A neuro-immune network for solving the traveling salesman problem. In: The 2006 IEEE International Joint Conference on Neural Network Proceedings. IEEE; 2006. pp. 3760–3766

[23] Saenphon T, Phimoltares S, Lursinsap C. Combining new fast opposite gradient search with ant colony optimization for solving travelling salesman problem. Engineering Applications of Artificial Intelligence. 2014;**35**:324-334

[24] Tuba M, Jovanovic R. Improved ACO algorithm with pheromone correction strategy for the traveling salesman problem. International Journal of Computers Communications & Control. 2013;**8**(3):477-485

[25] Yun H-Y, Jeong S-J, Kim K-S. Advanced harmony search with ant colony optimization for solving the traveling salesman problem. Journal of Applied Mathematics. 2013;**2013**

[26] Mehmet Ali Mustafa K, Faouzi Kamoun. Neural networks for shortest path computation and routing in computer networks. IEEE Transactions on Neural Networks. 1993;**4**(6):941–954

[27] Creput J-C, Koukam A. A memetic neural network for the euclidean traveling salesman problem. Neurocomputing. 2009;**72**(4–6):1250-1264

[28] Masutti TAS, de Castro LN. A self-organizing neural network using ideas from the immune system to solve the traveling salesman problem. Information Sciences. 2009;**179**(10):1454-1468

[29] Papadimitriou CH. The adjacency relation on the traveling salesman polytope is np-complete. Mathematical Programming. 1978;**14**(1):312-324

[30] Somhom S, Modares A, Enkawa T. A self-organising model for the travelling salesman problem. Journal of the Operational Research Society. 1997;**48**(9):919-928

[31] Marinakis Y, Marinaki M. A hybrid multi-swarm particle swarm optimization algorithm for the probabilistic traveling salesman problem. Computers & Operations Research. 2010;**37**(3):432-442

[32] Shi XH, et al. Particle swarm optimization-based algorithms for TSP and generalized TSP. Information Processing Letters. 2007;**103**(5):169-176

[33] Chen Y, Zhang P. Optimized annealing of traveling salesman problem from the nth-nearest-neighbor distribution. Physica A: Statistical Mechanics and Its Applications. 2006;**371**(2):627-632

[34] Roy GG et al. Design of non-uniform circular antenna arrays using a modified invasive weed optimization algorithm. IEEE Transactions on Antennas and Propagation. 2010;**59**(1):110-118

[35] Sengupta A, et al. Energy efficient trajectory planning by a robot arm using invasive weed optimization technique.

In: 2011 Third World Congress on Nature and Biologically Inspired Computing. IEEE; 2011. pp. 311-316

[36] Albayrak M, Allahverdi N. Development a new mutation operator to solve the traveling salesman problem by aid of genetic algorithms. Expert Systems with Applications. 2011;**38**(3): 1313-1320

[37] Freisleben B, Merz P. A genetic local search algorithm for solving symmetric and asymmetric traveling salesman problems. In: Proceedings of IEEE International Conference on Evolutionary Computation. IEEE; 1996. pp. 616-621

[38] Freisleben B, Merz P. New genetic local search operators for the traveling salesman problem. In: International Conference on Parallel Problem Solving from Nature. Springer; 1996. pp. 890-899

[39] Syswerda Gilbert. Scheduling Optimization Using Genetic Algorithms: Handbook of Genetic Algorithms. 1991

[40] Croes GA. A method for solving traveling-salesman problems. Operations Research. 1958;**6**(6):791-812

[41] Gutin G, Punnen AP. The Traveling Salesman Problem and its Variations. Springer Science & Business Media; 2002

[42] Zhang W. Truncated branch-and-bound: A case study on the asymmetric TSP. In: Proc. Of AAAI 1993 Spring Symposium on AI and NP-hard problems. Vol. 160166. 1993

[43] Zhang W. Depth-first branch-and-bound versus local search: A case study. In: AAAI/IAAI. 2000. pp. 930-935

[44] Karp RM. A patching algorithm for the nonsymmetric traveling-salesman problem. SIAM Journal on Computing. 1979;**8**(4):561-573

[45] Goldberg DE. Genetic algorithms in search. Optimization, and Machine Learning. 1989

[46] Davis L. Hybridization and numerical representation. In: The Handbook of Genetic Algorithms. 1991. pp. 61-71

[47] Oliver IM, Smith D, Holland JRC. Study of permutation crossover operators on the traveling salesman problem. In: Genetic Algorithms and Their Applications: Proceedings of the Second International Conference on Genetic Algorithms. Hillsdale, NJ: L. Erlhaum Associates; 1987

[48] Gunduz M, Kiran MS, Ozceylan E. A hierarchic approach based on swarm intelligence to solve the traveling salesman problem. Turkish Journal of Electrical Engineering & Computer Sciences. 2015;**23**(1):103-117

Section 3

Engineering Applications

Chapter 4

Programming an Evolutionary Algorithm for the Estimation of Non-Linear Damping Vibration Parameters

Carlos A. Lara and Cesar Guerra

Abstract

The use of genetic algorithms (GAs) has branched out into various disciplines such as mechanical engineering, providing solutions in cases where some models do not have a mathematical solution. In the field of mechanical vibrations, there are empirical nonlinear models that seek to represent the physical behavior of certain elements, such as in aeronautical applications, where stiffness and damping in structures can present hysteresis and can be represented by means of the Bouc-Wen (BW) model. This model includes constants that define the behavior of non-linear stiffness and damping, which are difficult to obtain since they are empirical models. This work presents the results of programming a GA to estimate the BW model constants for wire rope springs, commonly used as vibration isolators that have nonlinear stiffness and damping resulting in hysteresis behavior.

Keywords: genetic algorithms, Bouc-Wen model, non-linear stiffness, damping, rope springs

1. Introduction

The study of vibration control has taken great importance and necessity in recent years. The emergence of new isolating and vibration-dissipating elements has prompted new studies about their behavior in hysteresis damping. Damping is difficult to model this is often due to more than one phenomenon, for example, a combination of viscous damping, internal damping, dry friction, viscoelastic effects, etc. [1]. There are various of insulator configurations, among which are: metal springs, i.e., helical or leaf springs, and viscoelastic elements such as rubber, neoprene, silicon, air springs, etc. Steel cable springs are characterized by their high energy storage and dissipation capacity, based on dry friction [2].

In the study of hysteresis damping, there is a problem in describing its non-linear behavior [1]. One solution is the model of BW [3], from which heuristic algorithm techniques have been proposed, which seek to estimate the parameters, factors, error reduction, among others, of which it is possible to establish an estimated model of the system, and from it, design control and/or estimation strategies [4]. Given the non-linear nature of the model, it has been approached by different methods, including the following: Gauss–Newton, modified Gauss–Newton, Least squares, Simplex, Levenberg–Marquardt, extended Kalman filter, reduced gradient methods, genetic algorithms (GAs), real-coded GAs, Differential Evolution, adaptive laws, etc.

It is well known that the classical linearized analysis of the dynamical systems can lead to results that are reasonably accurate only when the minimum (rest position) force and the displacements are of such magnitude that the relative change in force during the motion is small. In practice, however, very often some or all of these assumptions are violated, so that in many dynamical systems nonlinear phenomena may completely alter intuitively expected behavior and can drastically change their dynamical responses [5].

In mechanical vibrations systems, the nonlinear phenomenon can be presented principally in the springs elements and/or the dampers models, significant results have also been obtained to represent these phenomena.

Recently, the use of new insulator mechanical in several systems, for example in Aeronautics, has prompted research to design new non-linear model representatives of this elements, where a memory-dependent, multivalued relation between force and deformation, i.e., hysteresis, is often observed in structural materials and elements, such as reinforced concrete, steel, base isolators, dampers, and soil profiles.

Many mathematical models have been developed to efficiently describe such behavior for use in time history and random vibration analyses. One of the most popular is the BW class of hysteresis which is used to describe the hysteretic behaviors of structures in nonlinear dynamic and stochastic analyses.

BW model is used to describe the non-linear behavior of the stiffness and damping of an element, where the restoring force becomes highly nonlinear showing significant hysteresis. The hereditary nature of this nonlinear restoring force indicates that the force cannot be described as a function of the instantaneous displacement and velocity. Accordingly, many hysteretic restoring force models were developed to include the time-dependent nature using a set of differential equations.

BW model is a semi-empirical model that contains several parameters and is one of the most used hysteretic models, and it was introduced by Robert Bouc [6] and extended by Yi-Kwei Wen [7] who demonstrated its versatility by producing a variety of hysteretic patterns.

Being a semi-empirical model, the BW model contains semi-empirical parameters which should be esteemed using several mathematical and empirical strategies, such as Gauss–Newton [8], modified Gauss–Newton [9], Least squares [10], Simplex [11], Levenberg–Marquardt [12], extended Kalman filters [13] among others.

Recently, the use of GAs for the estimation of BW parameters has been used, for example, Kwok et al. [14] used a GAs to estimate the parameters of the BW model with characteristics of non-symmetrical hysteresis; Wang et al. [15] used a novel differential evolution algorithm for estimation of parameters of asymmetric hysteresis loops.

Meanwhile, Charalampakis et al. [16] presented a new identification method that determines the parameters of Bouc-Web hysteresis based on a hybrid evolutionary algorithm which utilizes selected stochastic operators.

In most cases, the problem lies in that the BW model can be easily solved because it combines an algebraic equation with a differential equation, in addition, it is found that there are redundant parameters [17]. One solution to deal with this problem, users of the BW model often fix some parameters to arbitrary values, while other users eliminate the redundant parameters via a process of normalization.

In this work, a discrete approximation of the BW model is proposed to facilitate the estimation of BW parameters using an efficient evolutionary algorithm called "Evonorm". The programming of algorithm helped model the behavior of physically loaded/unloaded springs in an experimental setup.

2. BW discrete model approximated

Starting from a mass-spring-damper vibratory system, where the interest is based in accordance with the type of damping present (hysteresis or a combination of phenomena) as shown in **Figure 1**, the application of BW for the study of its behavior in terms of the damping present in the system.

Mass (m) represents the inertia in *kg* and Damper (c) is the viscous damping in *N*s/m*; both elements are considered as lineal elements of the system. Now, the spring (k) represents the stiffness lineal in N/m, but in our study, this element contains both the non-linear stiffness and non-linear damping.

Actuality, no any mathematical representation exists for this non-lineal phenomena, but an empirical representation known as BW model and are described in the following.

The BW model was proposed initially by Bouc early in 1967 and subsequently generalized by Wen in 1976. Its typical equations are expressed as follows:

$$R = k_e x + k_h z \quad (1)$$

$$\dot{z} = A\dot{x} - \beta|\dot{x}||z|^{n-1}z - \gamma\dot{x}|z|^n \quad (2)$$

Where R is the restoring force, $x = x(t)$ is the deformation of spring, $z = z(t)$ is known as hysteresis displacement and not physically measured, $\dot{x} = dx/dt$ and $\dot{z} = dz/dt$; $k_e = \alpha k$ and $k_h = (1 - \alpha)k$; where k is the lineal stiffness coefficient.

The α parameter is the ratio of post-yield to pre-yield stiffness, A, n, β, γ are parameters that control the hysteresis shape.

Analyzing Eqs. (1) and (2), it is found that there are redundant parameters in BW model. That is why to estimate the control parameters in the BW model, have been defined with alternative models or modified; for example: "The normalized BW model" presented in [18], "A multi-objective optimization algorithm for Bouc–Wen–Baber–Noori model [19].

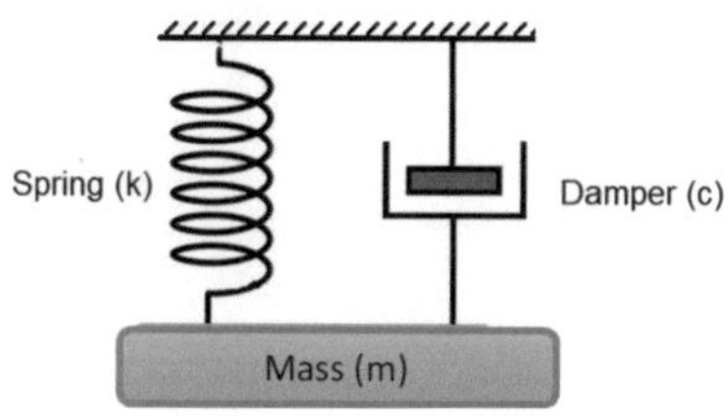

Figure 1.
m-k-c system.

In this work, the first step in order to solve the BW equation consists of eliminating the derivative function using a numerical approximation, in this case discrete approximation of the first-order of the original model called "Euler-backward discretization" is used as:

$$\dot{x} \approx \frac{x_{k+1} - x_k}{\Delta t} \tag{3}$$

Where x_{k+1}, x_k are the two-sample data of the x(t) and Δt is the time sample of the sample data width as shown in **Figure 2**.

Now, from Eq. (1), it is possible to define the differential equation $\dot{R} = k_e\dot{x} + k_h\dot{z}$, therefore applying the discrete approximation (3) to $\dot{R}, \dot{x}$ and $\dot{z}$ the following equation is applied.

$$R_{k+1} - R_k = k_e(x_{k+1} - x_k) + k_h(z_{k+1} - z_k) \tag{4}$$

$$z_{k+1} - z_k = A(x_{k+1} - x_k) - \beta|x_{k+1} - x_k||z_k|^{n-1}z_k - \gamma(x_{k+1} - x_k)|z_k|^n \tag{5}$$

This model can be solved in programming if the deformation $\Delta x = x_{k+1} - x_k$ is fixed or $\Delta R = R_{k+1} - R_k$ is fixed, is clear that if Δx is fixed, so ΔR is easy to determine. In the sample, *k = 0* is clear that $x_0 = 0$, $R_0 = 0$ and $z_0 = 0$, now the next values can be calculated using the next Pseudocode as shown in **Table 1**.

1. $\Delta x = x_{k+1} - x_k$ is fixed and given.
1. $A, \beta, \gamma, n, \alpha$ and k is given.
1. In *k* = *0*, $z_k = 0$, $R_k = 0$.
1. Determine z_{k+1} using (5).
1. Determine R_{k+1} using (4).
1. Set $z_k = z_{k+1}$, $R_k = R_{k+1}$
1. Go to (4) to calculate the values of the next sample.

Table 1.
Pseudocode to solve the BW discrete model.

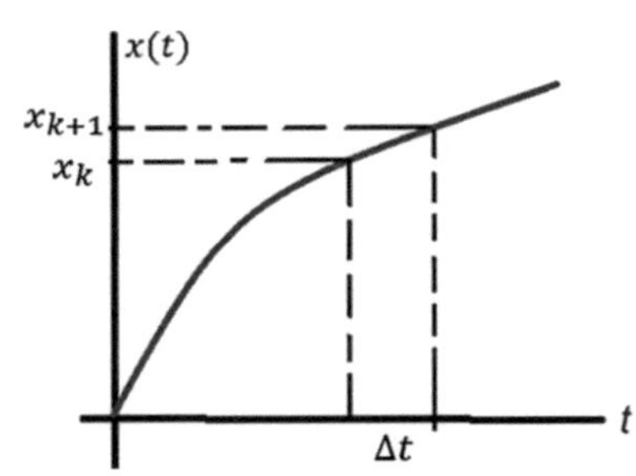

Figure 2.
Euler backward approximation.

3. Evolutionary algorithm model

Evonorm [20] is an evolutionary algorithm used in this work to estimate the parameters of BW model. Next definitions are required to understand the algorithm.

3.1 Variables definition

Decision variables (Y). These are the m-variables of the system with unknown values. The determination of these values will be the target of the algorithm.

$$Y \equiv \{y_1, y_2, y_3, \dots, y_m\} \tag{6}$$

Design variables (X). These are the n-variables of the system with known values, these values set that will allow determinate the decision values (Y) using the algorithm.

$$X \equiv \{x_1, x_2, x_3, \dots, x_n\} \tag{7}$$

Decision values $(\tilde{Y})$. These are the values of Decision variables (Y).

$$\tilde{Y} = [\tilde{y}_1 \ \tilde{y}_2 \ \cdots \ \tilde{y}_m] \tag{8}$$

Design values $(\tilde{X})$. These are the values of Design variables that will be used to determine the decision values in the algorithm. The efficiency of the algorithm depends on selecting the p-samples required, therefore $\tilde{X}$ is a matrix of $p \times n$ dimension

$$\tilde{X} = \begin{bmatrix} \tilde{X}(1) \\ \tilde{X}(2) \\ \vdots \\ \tilde{X}(p) \end{bmatrix} = \begin{bmatrix} \tilde{x}_1(1) & \tilde{x}_2(1) & \cdots & \tilde{x}_n(1) \\ \tilde{x}_1(2) & \tilde{x}_2(2) & \cdots & \tilde{x}_n(2) \\ \vdots & \vdots & \ddots & \vdots \\ \tilde{x}_1(p) & \tilde{x}_2(p) & \cdots & \tilde{x}_n(p) \end{bmatrix} \tag{9}$$

Fitness function. It is a function or heuristic algorithm required to evaluate if the values of decision variables and design variables are correct, namely.

Minimal error $(\bar{e})$. It is a selected value such that the fitness function evaluation, allows to affirm that the $\tilde{Y}$ values are correct:

$$\sum_{i=1}^{p} \lfloor f(\tilde{X}(i), \tilde{Y}) \rfloor < \bar{e} \tag{10}$$

Population (P). It is a q-data set (individuals) of the Decision values candidates to solve (10), therefore P is a matrix of $q \times m$ dimension

$$P = \begin{bmatrix} \tilde{Y}(1) \\ \tilde{Y}(2) \\ \vdots \\ \tilde{Y}(q) \end{bmatrix} = \begin{bmatrix} \tilde{y}_1(1) & \tilde{y}_2(1) & \cdots & \tilde{y}_m(1) \\ \tilde{y}_1(2) & \tilde{y}_2(2) & \cdots & \tilde{y}_m(2) \\ \vdots & \vdots & \ddots & \vdots \\ \tilde{y}_1(q) & \tilde{y}_2(q) & & \tilde{y}_m(q) \end{bmatrix} \tag{11}$$

3.2 Selection (Ts)

Since each row of the P-matrix in (9) represents a possible solution of the condition (10), it is necessary determinate the value of the error for each row of the population matrix (11),

$$E(j) = \sum_{i=1}^{p} \lfloor f(\tilde{X}(i), \tilde{Y}(J)) \rfloor \tag{12}$$

Now, for each j-individual in the matrix population, it is necessary to evaluate the contribution it makes to the solution (12), for which the rows of the P-matrix (11), must be ordered in ascending order. Now, must be selected the fits Ts-individuals (rows) to mutate and crossover to generate the new Population.

3.3 Mutation and crossover

In order to avoid the algorithm being trapped in local optima, Evonorm uses random variables with normal distribution. The normal distribution function is a random variable and describes many random phenomena that occur in everyday life. It simulated the normal distribution function with two parameters, the first is the mean and it is a numeric measure of the central tendency of the random variable. The second parameter is the standard deviation, and it is a measure of the dispersion of a variable around the mean. A normal distribution function can be used to represent a set of possible values of a decision variable, so it is necessary to use a set of parameters (mean and standard deviation) of the normal distribution function per decision variable (18).

Therefore, the mutation is generated using each k-column of the (11) and to Ts-individuals select as follows:

$$\mu_k = \frac{\sum_{i=1}^{Ts} \tilde{y}_k(i)}{Ts} \sigma_k = \sqrt{\frac{\sum_{i=1}^{Ts} \left(\tilde{y}_k(i) - \mu_k\right)^2}{Ts}} \tag{13}$$

At the same time, the new population is generated as follows:

$$\tilde{y}_k(i) = \begin{cases} \mu_k + N(0,1)\sigma_k \text{ if } U(.) > 0.5 \\ \tilde{y}_k(i) + N(0,1)\sigma_k \text{ if } U(.) \leq 0.5 \end{cases} \tag{14}$$

Where $N(0,1) = \sum_{i=1}^{12} U(.) - 6$, and $U(.)$ is a random value between 0 and 1. The pseudocode of the Evonorm algorithm is shown in **Table 2**.

1. Generation of initial population (11).
2. Evaluation of initial population.
3. Selection of the best $Ts < q$ individuals.
4. Mutation and generation of new population (13) and (14).
5. If a criterion (10) satisfied, then end else go to step 2.

Table 2.
Pseudocode of Evonorm.

4. Experiment result

4.1 Experimental data

The experimental data were obtained from tests made to a set of four wire rope isolators (WRI) in parallel (like the one shown in **Figure 3**), to have stability during the test.

The mechanical tests were carried out in the universal machine Shimadzu (**Figure 4**) of 10 KN, with controlled displacement for compression-decompression load.

The values of Force-deformation in the wire ropes were plotted in **Figure 5**, showing the hysteretic behavior of the WRI.

Figure 3.
Picture and physical characteristics of the wire rope isolator (WRI) used in the experimental procedure.

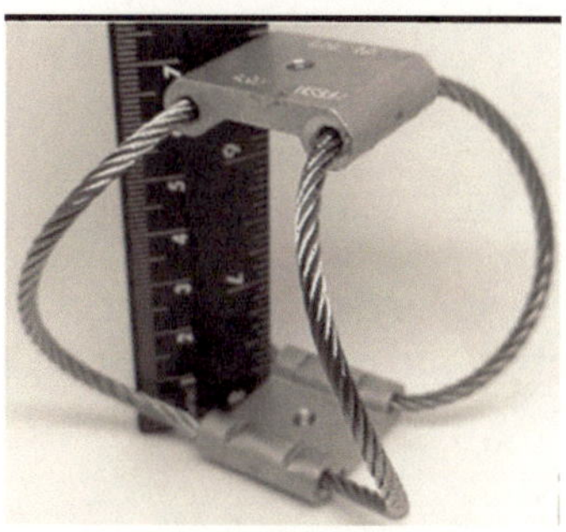

Figure 4.
Experimental setup for the static monotonic and cyclic tests.

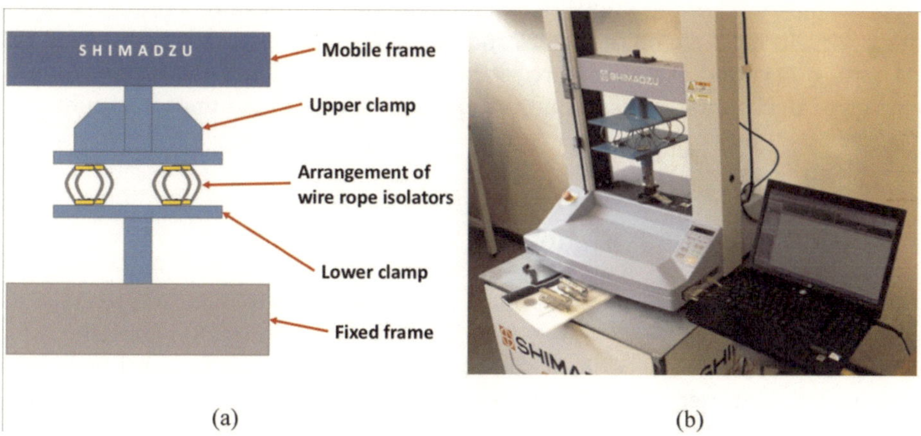

Figure 5.
Data result from experimentation.

From the data that are plotted in **Figure 5**, representative samples were taken, this is to input variables for the evolutionary algorithm, which performs the calculation of distribution function parameters to generate new individuals.

4.2 Algorithm programming

In order to apply the Evonorm algorithm to estimate the BW parameters model, which is necessary to define the decision variables and design variables of the system. In the BW model the data set of hysteresis loop are the design variables, Meanwhile the parameters $\alpha, \beta, \gamma, n, A, k$ are the decision variables:

$$Y \equiv \{\alpha, \beta, \gamma, n, A, k\} \tag{15}$$

$$X \equiv \{x_{k+1}, x_k, R_{k+1}, R_k\} \tag{16}$$

The values of the design variables can be obtained from the result experiment with the values in **Figure 5**. In this case, the values that appear below are selected::

$$X_v = \begin{bmatrix} -1.53 & -1.51 & -10.37 & -10.32 \\ -1.17 & -1.28 & -4.61 & -5.93 \\ -1.04 & -1.17 & -3.30 & -4.61 \\ -0.20 & -0.39 & 2.66 & 1.62 \\ 0.70 & 0.53 & 7.32 & 6.57 \\ 1.36 & 1.27 & 0.51 & 9.21 \\ 1.52 & 1.64 & 10.04 & 10.19 \end{bmatrix} \tag{17}$$

Now, it is necessary to determine the fitness function $F(Xd, Yd)$ in the algorithm, this function included the design variables Xd and de decision variables Yd. Therefore, the fitness function is the approximated model discrete (6).

$$F(X, Y) = R_{k+1} - R_k - k_e \Delta_X k_h \left(A\Delta_X + \beta|\Delta_X||z_k|^{n-1} z_k \gamma + \Delta_X |z_k|^n \right) \tag{18}$$

The major objective will be to minimize the function $F(Xd, Yd) \cong 0$.

Next values was used in the algorithm programming: Number of individuals: p = 50, number of selected individuals: *Ts = 25* and the numbers of iterations: (iters) *Nr* = 100.

On the other hand, it is necessary establish limits in the Design variables, this provides to algorithm's heuristic find the optimal values and in minor iterations.

The limits values of the Design variables are:

$$\alpha\epsilon[0, 1], \beta\epsilon[0.1,0.9], \gamma\epsilon[0.1,0.9], \tag{19}$$

$$A\epsilon[1, 10], n\epsilon(1, 2], k\epsilon[10, 1000] \tag{20}$$

The selection of these values and the number of samples is important to make the algorithm run more efficient, so they must be strategic and minority. Strategic because, for this case, the middle of the loop was selected, taking as samples: the ends, the point at the intersection with the vertical, as well as two intermediate points at the ends of the intersection with the vertical; and minorities so as not to increase the computational cost of the algorithm. The result obtained is shown in **Table 3** for different runs.

α	A	N	β	γ	k	Error
0.40	1.69	1.1	0.81	0.72	7.78	0.21
0.43	1.35	1.1	0.78	0.79	8.53	0.23
0.53	1.98	1.1	0.86	0.64	6.96	0.23
0.46	1.50	1.1	1.12	0.37	7.92	0.25

Table 3.
Evonorm algorithm results.

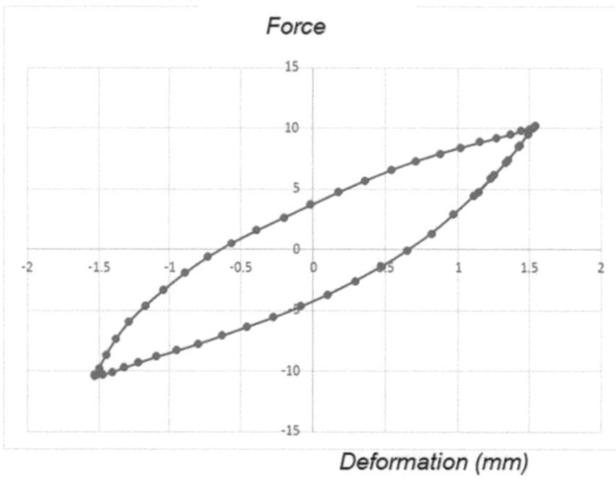

Figure 6.
Error graph.

Table 3 shows the results of the Evonorm algorithm with the values that define the shape of the hysteresis loop.

The convergence of the error showing the efficiency of the algorithm is also shown. **Figure 6** shows the graph of the error percentage against the number of iterations performed by the algorithm.

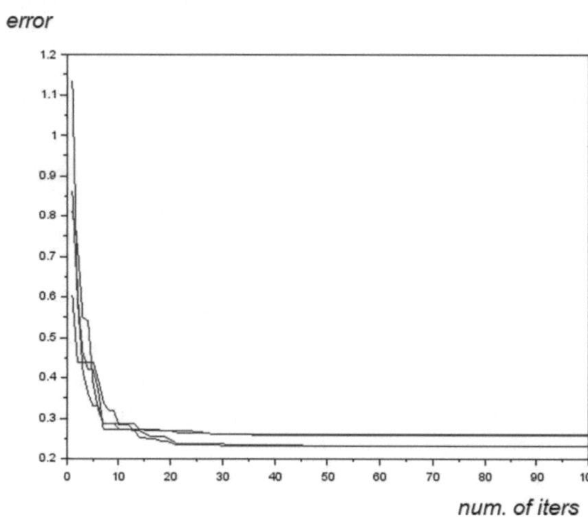

Figure 7.
Comparison between real graph and loops obtained from evolutionary algorithm.

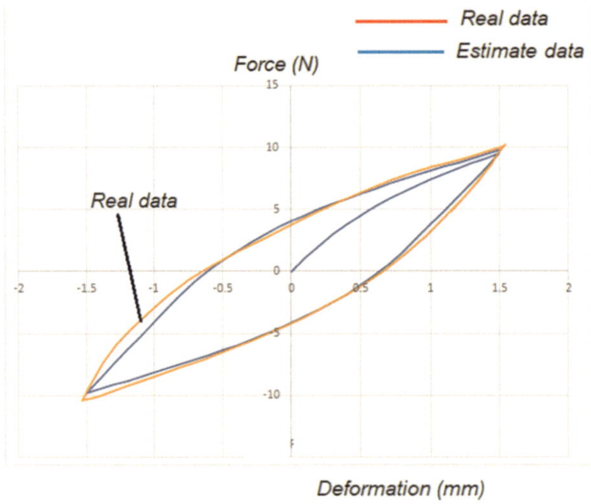

Figure 8.
Graph of actual data vs estimated data

On the other hand, the real results obtained from experimental tests (represented by the red graph) and the results with the BW parameters obtained from the evolutionary algorithm (represented by the blue graph) were graphed (**Figure** 7). The data that was chosen to feed the BW model were those that were obtained with the minimum percentage of error. **Figure 8** shows the comparison of the real values versus those produced by the evolutionary algorithm.

5. Conclusion and future work

The evolutionary algorithm can obtain adjusted BW parameters that can be fed into the model. The experimental hysteresis loops were fitted, with the parameter from Evonorm, in only 4 runs of the evolutionary algorithm.

The convergence of the minimum value of the error of the different permutations was achieved in 25 iterations, which is acceptable and can be improved by feeding more amount of experimental data to the evolutionary algorithm.

Of the 4 runs that were carried out with the Evonorm evolutionary algorithm, very similar ranges of errors were obtained. With the best option, very similar graphs, of the real experimentation and of the algorithm, were also obtained, and for that reason can be confident with the values that the algorithm delivers.

From **Figure** 7, the parameters provided by the EVONORM evolutionary algorithm, which are fed to the BW model, generate a graph that is very close to the graph that is made with the values obtained from experimental tests. Given the above, performing the selection of the input variables to the evolutionary algorithm correctly and strategically makes the estimated output values that are fed to the model guaranteed precision to the hysteresis loop compared to that produced by experimental tests.

It is clear that the correct selection of limits values in Evonorm is a condition to ease the convergence, these values can be obtained using the knowledge of the expert;

but is possible to use a neural network, so that after the training, estimate the values of the limits, which a future work.

Finally, concerning the error value, in **Figure 6** this remains constant after 25 iterations and the ideality is that this value declines gradually, some changes in the limits values can tackle this, furthermore, it is possible to improve tunning the percentage values in Eq. (14), for example, if $U(.) > 0.4$, if $U(.) \leq 0.4$.

Author details

Carlos A. Lara* and Cesar Guerra
Universidad Autónoma de Nuevo León, San Nicolás de los Garza, México

*Address all correspondence to: carlos.laraoc@gmail.com

References

[1] Rao SS. Mechanical Vibrations. 6th ed. Pearson; 2016. p. 1152

[2] Ledezma D, Tapia P, Ferguson N, Brennan M, Tang B. Recent advances in shock vibration isolation: An overview and future possibilities. ASME. Applied Mechanics Reviews. 2019;**71**(6):060802. DOI: 10.1115/1.4044190

[3] de Silva CW. Vibration Damping, Control, and Design. Press; 2019. p. 634

[4] Hassani V, Tjahjowidodo T, Do TN. A survey on hysteresis modeling, identification and control. Mechanical Systems and Signal Processing. 2014;**49** (1–2):209-233. DOI: 10.1016/J.YMSSP. 2014.04.012

[5] Warminski J, Lenci S, Carttmell P, Rega G, Wiercigroch M. Nonlinear Dynamic Phenomena, Mechanical Engineering Systems in Solids Mechanics and Its Applications. 2012. p. 181

[6] Bouc R. Forced vibration of mechanical systems with hysteresis. In: Proceedings of the Fourth Conference on Nonlinear Oscillation. Prague, Czechoslovakia; 1967. p. 315

[7] Wen YK. Method for random vibration of hysteretic systems. Journal of Engineering Mechanics. American Society of Civil Engineers. 1976;**102**(2): 249-263

[8] Yar M, Hammond J. Parameter estimation for hysteretic systems. Journal of Sound and Vibration. 1987; **117**(1):161-172. DOI: 10.1016/0022-460X (87)90442-1

[9] Kunnath S, Mander J, Fang L. Parameter identification for degrading and pinched hysteretic structural concrete systems. Engineering Structures. 1997;**19**(3):224-232. DOI: 10.1016/S0141-0296(96)00058-2

[10] Sues R, Mau S, Wen Y. System identification of degrading hysteretic restoring forces. Journal of Engineering Mechanics. 1988;**114**(5):833-846. DOI: 10.1061/(ASCE)0733-9399(1988) 114:5(833)

[11] Zhang H, Foliente G, Yang Y, Ma F. Parameter identification of inelastic structures under dynamic loads. Earthquake Engineering and Structural Dynamics. 2002;**31**:1113-1130. DOI: 10.1002/eqe.151

[12] Ni Y, Ko J, Wong C. Identification of non-linear hysteretic isolators from periodic vibration tests. Journal of Sound and Vibration. 1998;**217**(4):737-756. DOI: 10.1006/jsvi.1998.1804

[13] Lin J, Zhang Y. Nonlinear structural identification using extended Kalman filter. Computers and Structures. 1994; **52**(4):757-764. DOI: 10.1016/0045-7949 (94)90357-3

[14] Kwok N, Ha Q. Bouc-Wen model parameter identification for a MR fluid damper using computationally efficient FA. ISA Transaction. 2007;**46**:167-179. DOI: 10.1016/j.isatra.2006.08.005

[15] Wang G, Chen G, Bai F. Modeling and identification of asymmetric Bouc-Wen hysteresis for piezoelectric actuator via a novel differential evolution algorithm. Sensor and Actuators. 2015; **235**:105-118. DOI: 10.1016/j. sna.2015.09.043

[16] Charalampakis A, Koumousis V. Identification of Bluc-Wen hysteresis system by a hybrid evolutionary algorithm. Journal of Sound and

Vibration. 2008;**314**:571-185. DOI: 10.1016/j.jsv.2008.01.018

[17] Zhu X, Lu X. Parametric identification of Bouc-Wen model and its application in mild steel damper modeling. Procedia Engineering. 2011; **14**:318-324. DOI: 10.1016/j.proeng. 2011.07.039

[18] Zhu H, Rui X, Yang F, Zhu W, Wei M. An efficient parameters identification method of normalized Bouc-Wen model for MR damper. Journal of Sound and Vibration. 2019; **448**:146-158. DOI: 10.1016/j.jsv.2019. 02.019

[19] Li Z, Noori M, Zhao Y, Wan C, Feng D, Altabey WA. A multi-objective optimization algorithm for Bouc–Wen–Baber–Noori model to identify reinforced concrete columns failing in different modes. Proceedings of the Institution of Mechanical Engineers, Part L: Journal of Materials: Design and Applications. 2021;**235**(9):2165-2182. DOI: 10.1177/14644207211020028

[20] Torres L. Evonorm: Easy and effective implementation of estimation of distribution algorithms. Journal of Research in Computing, Science. 2008; **23**:75-83

Chapter 5

The Genetic Algorithm and its Application in Calculating the Kinetic Parameters of the Thermoluminescence Curve

Nguyen Duy Sang

Abstract

This chapter explores the use of genetic algorithms as a tool for calculating the kinetic parameters of the thermoluminescence curve. Genetic algorithm is a search algorithm inspired by the process of natural selection, and it has proven to be effective in solving optimization problems in various fields. Author used genetic algorithm to estimate the activation energy and frequency factor of the thermoluminescence curve, which are important parameters in determining the dosimetric properties of materials. The results showed that genetic algorithm could accurately estimate the kinetic parameters of the thermoluminescence curve with high precision and efficiency compared to conventional methods. This approach can also handle noisy data and reduce the impact of outliers on the estimation process. Furthermore, author demonstrated that genetic algorithm can be generalized to different types of the thermoluminescence curves, such as those generated by different materials or under different experimental conditions. The proposed method is fast, accurate, and robust, making it useful for researchers in the field of dosimetry who require precise estimations of these parameters.

Keywords: genetic algorithms, kinetic parameters, thermoluminescence, materials, dosimetry

1. Introduction

The study of thermoluminescence (TL) has been an important area of research for many years due to its numerous applications in dating, dosimetry, and material characterization. The TL curve is a graph of the light emitted by a material as it is heated, and it provides valuable information about the energy states of the material. In order to extract useful information from the TL curve, it is necessary to analyze its kinetic parameters, such as activation energy and frequency factor. One powerful method for calculating these parameters is the genetic algorithm, which is a computational technique based on the principles of natural selection and evolution. Genetic algorithm is based on the simulation of genetic processes in living organisms and the principle of natural evolution. The experimentally obtained TL spectra are curves that are fitted according to different kinetic models [1–3]. The algorithm works with a biological population, each of which represents the ability to adapt to the explore space through synchronous combinations of evolutionary processes such as selection, crossover, and mutation [4].

Python provides a powerful platform for TL analysis, allowing for complex data analysis and visualization for accurate interpretation of experimental results. Python can be used for TL analysis by performing data analysis on experimental data obtained from TL measurements.

Here are some steps that can be followed: Import necessary modules: The first step is to import the necessary modules that will be required for data analysis. Some commonly used modules include NumPy, Pandas, Matplotlib, and Seaborn. Load experimental data: Load the experimental data into Python using the Pandas module. The data should be stored in a CSV or TXT file format that can be easily loaded into Python. Data processing: The next step is to clean and process the data. This involves removing any noise or unwanted signals present in the data. This can be done using NumPy's signal processing functions such as filtering and smoothing. Plotting data: Visualizing the processed data is an important step in TL analysis. Using Matplotlib and Seaborn, create different types of plots such as scatter plots, line plots, histograms. Curve fitting: The next step is to fit a curve to the data. This helps in determining the kinetic parameters of the material being studied. Python provides various curve fitting algorithms such as least squares fitting and nonlinear regression. Analyzing results: Finally, interpret the results obtained from curve fitting and apply them to the material being studied.

2. Genetic algorithm approach to the TL curves

2.1 Theoretical basis

Genetic algorithm (GA) is based on the simulation of genetic processes in living organisms and the principle of natural evolution. The algorithm works with a biological population, each of which represents the ability to adapt to the explore space through synchronous combinations of evolutionary processes such as selection, crossover, and mutation [4]. The GA tuning process includes the following steps: (i) Setting initial chromosome and its encoding; (ii) Calculate GOK model distributions for each variable from individuals of a population and evaluate the fit of the fitness function; (iii) Select randomly parents and go to the next step; (iv) Crossover and mutation, go to the next step; (v) Select randomly individuals and go to the next step; (vi) Accept the results if there is better fitness value than the worst explore in the population and go to the next step, reject the worst explore and return step (iii); (vii) If the number of pre-determined steps (stopping condition) is reached and go to the next step; and (viii) Print results for best explore in population and GA finish. The block diagram of the GA is shown in **Figure 1**.

Explanation of **Figure 1**: **Initial population**: create an initial population of candidate solutions randomly. Each solution represents a potential solution to the problem being solved. Selection: The selection process is typically based on a **fitness function** that evaluates each individual's performance. **Crossover:** Two selected individuals are

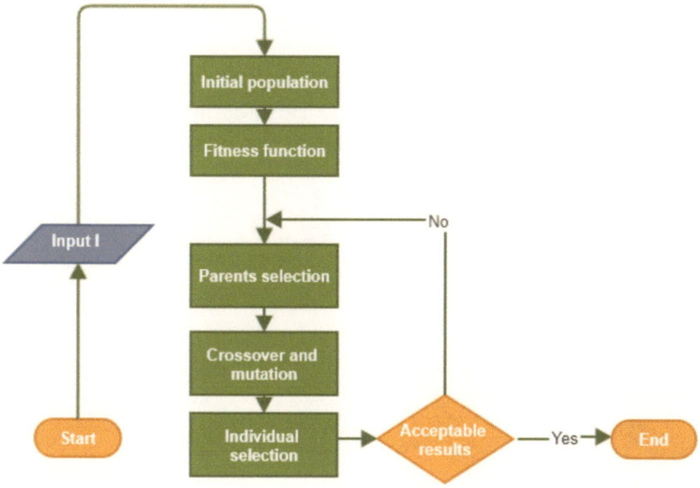

Figure 1.
Flowchart of the genetic algorithm.

combined to produce offspring with characteristics from both parents. The crossover point is chosen randomly, and the resulting offspring replace their parents in the population. **Mutation**: To maintain diversity in the population, some individuals undergo random mutation, which introduces new characteristics not present in the parent population. **Evaluation**: The fitness function is used to evaluate the performance of each individual in the population, including the newly generated offspring. **Termination**: The algorithm terminates when either a termination criterion is met (e.g., a maximum number of generations) or the best solution has been found.

Code in Python:

```
#Individual class
class Individual:
    def __init__(self, gene_list: List[float]) -> None:
        self.gene_list = gene_list
        self.fitness = func(self.gene_list[0])
    def get_gene(self):
        return self.gene_list[0]
    def crossover(cls, parent1, parent2):
        child1_gene,
    child2_gene = crossover_blend(parent1.get_gene(),
        parent2.get_gene(), 1, -10, 10)
    return      Individual([child1_gene]),Individual([child2_gene
])
    def mutate(cls, ind):
        mutated_gene =
    mutate_gaussian(ind.get_gene(), 0, 1, -10, 10)
        return Individual([mutated_gene])
    def select(cls, population):
        return select_tournament(population,tournament_size=3)
    def create_random(cls):
        return Individual([random.randrange(0, 900)])
```

The Python file FSG_GA.py at GitHub [5].

2.2 The thermoluminescence kinetic equation

The most commonly used models for analyzing such signals are the first order kinetics (FOK), second order kinetics (SOK), the general order kinetics (GOK) [6, 7].

The empirical equation describing the first-order TL curve has the form:

$$I(T) = I_m \exp\left[1 + \frac{E}{kT}\frac{T - T_m}{T_m} - \frac{T^2}{T_m^2}\exp\left(\frac{E}{kT}\frac{T - T_m}{T_m}\right)\left(1 - \frac{2kT}{E}\right) - \frac{2kT_m}{E}\right] \tag{1}$$

Code in Python:

```
def func(x):
    a,b,c=100, 473, 0.35
    kbz = 8.617385e-5
    return a*np.exp(1.0+c/kbz/x*((x-b)/b) -
((x/b)**2)*np.exp(c/kbz/x*((x-b)/b*(1.0-2.0*kbz*x/c)-
2.0*kbz*b/c)
```

Graph of first-order kinetic TL curve:

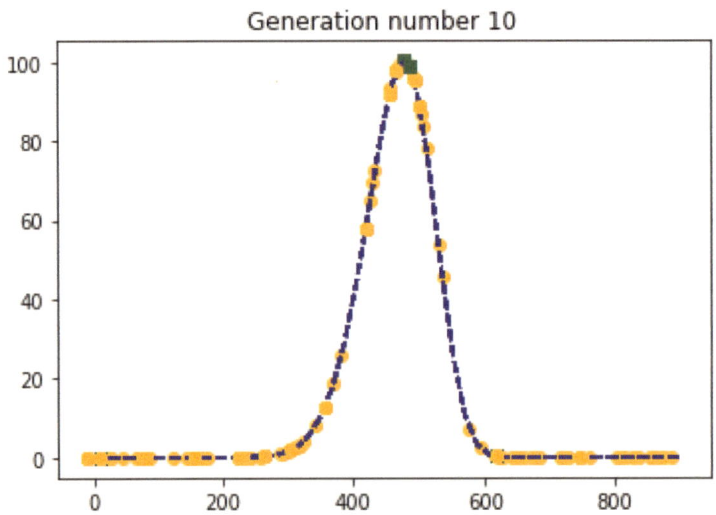

Figure 2.
First-order glow peaks of TL curve.

Explanation of **Figure 2**: Graph of the first-order kinetic model of the TL spectrum with a peak and symmetry.

The empirical equation describing the second-order TL curve has the form:

$$I(T) = 4I_m \exp\left(\frac{E}{kT}\frac{T - T_m}{T_m}\right)\left[\frac{T^2}{T_m^2}\left(1 - \frac{2kT}{E}\right)\exp\left(\frac{E}{kT}\frac{T - T_m}{T_m}\right) + 1 + \frac{2kT_m}{E}\right]^{-2} \tag{2}$$

Code in Python:

```
def func(x):
    a,b,c=100, 473, 0.35
    kbz = 8.617385e-5
    return 4*a*np.exp(c/kbz/x*((x-b)/b))
*(((x/b)**2)*np.exp(c/kbz/x*((x-b)/b)) *(1.0-
2.0*kbz*x/c)+1+2.0*kbz*b/c)**(-2)
```

Graph of second-order kinetic TL curve:

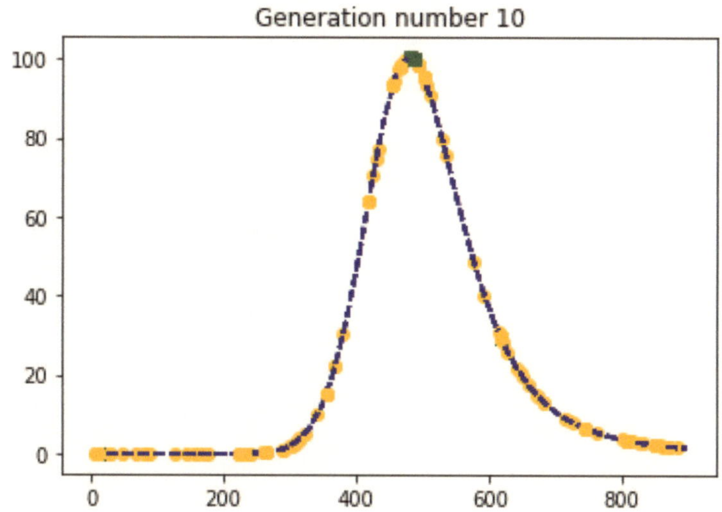

Figure 3.
Second-order glow peaks of TL curve.

Explanation of **Figure 3**: A graph of the second-order kinetic model of the TL spectrum with a peak and an asymmetrical shape, sloped forward of the peak.

General-order glow peaks are produced in intermediate cases (neither of first-order nor of second-order). The four parameters describing a glow peak are I_m, E, T_m, and b.

The empirical equation describing the general-order TL curve has the form:

$$I(T) = I_m b^{\frac{b}{b-1}} \exp\left(\frac{E}{kT}\frac{T-T_m}{T_m}\right)\left[(b-1)\left(1-\frac{2kT}{E}\right)\frac{T^2}{T_m^2}\exp\left(\frac{E}{kT}\frac{T-T_m}{T_m}\right)+1 + \frac{2kT_m(b-1)}{E}\right]^{-\frac{b}{b-1}} \quad (3)$$

Code in Python:

```
def func(x):
    a,b,c,bv = 100,473,0.35,1.5
    kbz = 8.617385e-5
    return a*(bv**(bv/(bv-1.0)))*np.exp(c/kbz/x*((x-b)/b))*(((bv-
1.0)*(x/b)**2)*np.exp(c/kbz/x*((x-b)/b))*(1.0-2.0*kbz*x/c)+1+(bv-
1.0)*2.0*kbz*b/c)**(-bv/(bv-1.0))
```

Graph of general-order kinetic TL curve:

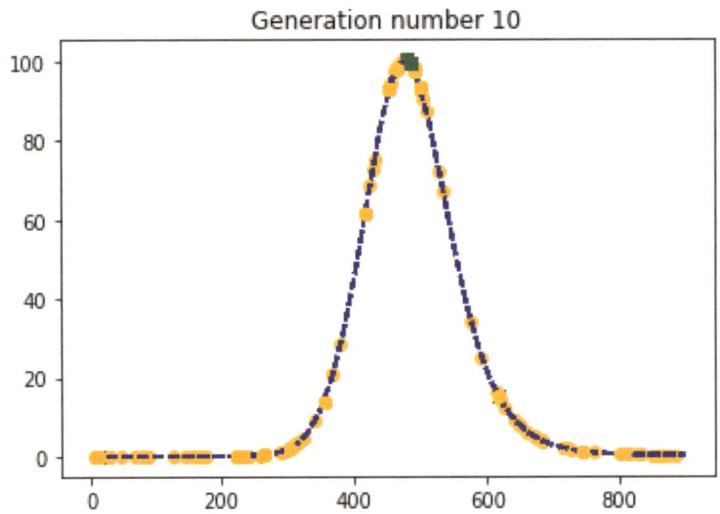

Figure 4.
General-order glow peaks of TL curve.

Explanation of **Figure 4**: The graph of the general-order kinetic model of the TL spectrum has the form of an intermediate peak of a first- and second-order kinetics model, with a slightly sloping front peak.

In addition to the three kinetic models above, the TL spectrum is also matched according to the mixing model and the one-trap one-center recombination model. Each model has its own advantages and disadvantages in calculating TL trap parameters.

Details and full Python source code for TL kinetic models can be found on GitHub [5].

2.3 Fitting TL curves to estimate the energy value

2.3.1 Straight line fitting

The initial rise (IR) method introduced by Garlick and Gibson [8] is used to estimate the trapping energy value of the TL curve. The IR method works as follows: A sample is irradiated with a known dose of ionizing radiation. The sample is then heated at a constant rate, and the emitted light is measured as a function of temperature. This is called a TL curve. The TL curve is divided into equal temperature intervals, and the total integrated light output for each interval is calculated. The integrated light output is plotted against the natural logarithm of the heating rate for each interval. The slope of the resulting straight line is proportional to the activation energy required to release the trapped charges in the sample. This method is based on the principle that the intensity of TL increases initially with the temperature. The activation energy can be calculated using the Arrhenius Eq. (4)

$$I \propto \exp\left(-\frac{E}{kT}\right) \tag{4}$$

where E is the activation energy, k is the Boltzmann constant, and slope is the slope of the straight line. The IR method is repeated at different doses of ionizing radiation, and the activation energy is plotted against the dose. The trapping energy value can be estimated from the intercept of this plot with the x-axis (dose axis). In summary, the IR method involves measuring the total integrated light output as a function of temperature for a sample irradiated with a known dose of ionizing radiation [8–11]. An example of an IR region of a glow peak is shown in **Figure 5**.

Explanation of **Figure 5**: The low-temperature peak tail in this region increases up to a critical temperature T_C which is less than T_m. The values of E from the IR remain true for some critical values of temperature up to T_C, corresponding to a TL intensity I_C smaller than about 10% of the maximum TL intensity I_m [12].

Code in Python:

```python
def IR(x, E, b):
    return E * x + b
def AE_IR(x,a,b,c,bv):
    y=gaussian(x,a,b,c,bv)
    Im = geneticTLy(x,a,b,c,bv)
    j=10
    Tci=[]
    Ici=[]
    for i in range(1,j):
        yi=np.zeros_like(x)+Im*i/100
        idx = np.argwhere(np.diff(np.sign(y - yi))).flatten()
        Tc = x[idx][0]
        Ic = y[idx][0]
        Tci=np.append(Tci, Tc)
        Ici=np.append(Ici, Ic)
    kbz = 8.617385e-5
    x_ND=1/(kbz*Tci)
    y_ln=np.log(Ici)

    # curve fit
    popt, _ = curve_fit(IR, x_ND, y_ln)
    # summarize the parameter values
    E_IR, b_IR = popt
    print("E & b:",-E_IR,b_IR)
    print('y = %.5f * x + %.5f' % (E_IR,b_IR))
    print("E_IR=",-E_IR)

    #define function to calculate adjusted r-squared
    coeffs = np.polyfit(x_ND, y_ln, 1)
    p = np.poly1d(coeffs)
    yhat = p(x_ND)
    ybar = np.sum(y_ln)/len(y_ln)
    ssreg = np.sum((yhat-ybar)**2)
    sstot = np.sum((y_ln - ybar)**2)
    R2 = 1- (((1-(ssreg/sstot))*(len(y_ln)-1))/(len(y_ln)-1))
    print("R$**2$=",R2)
    return -E_IR
```

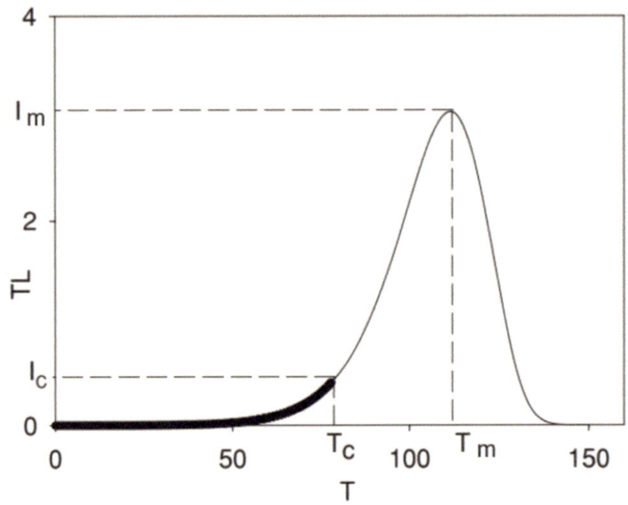

Figure 5.
Diagram of initial rise (IR).

Linear spectral fitting graph:

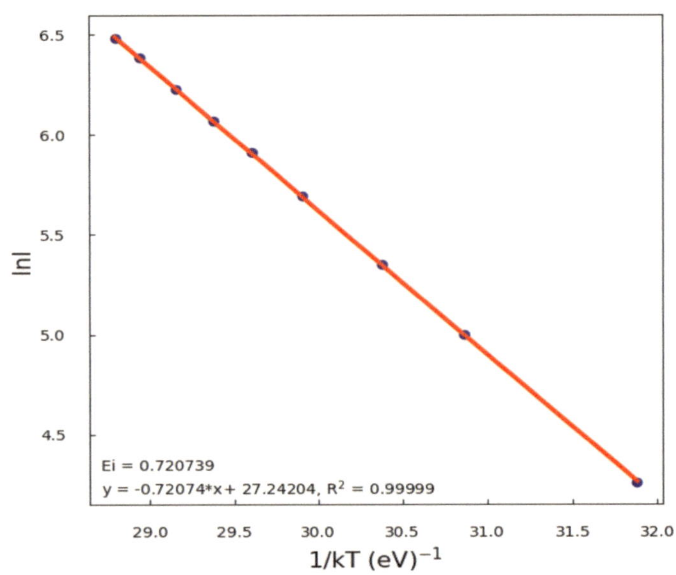

Figure 6.
Fitting line-fit of the TL1 to estimate E.

Explanation of **Figure 6**: On the left is the TL1 sample matched by GA and applying the IR method, the kinetic parameters are also calculated, E = 0.72 eV.

2.3.2 Gaussian peak spectral fitting

The peak shape (PS) method is generally called as Chen's [13] method, which is used to determine the kinetic parameters of the main glow peak of the TL materials. This

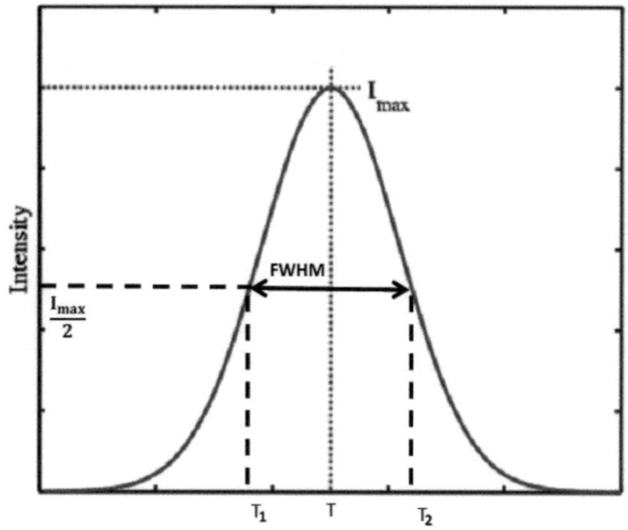

Figure 7.
Diagram of peak shape (PS).

method is mainly based on the temperatures T_m, T_1, and T_2, which are the peak temperatures, the temperatures at half of the maximum intensity, on the ascending and descending parts of the peak, respectively. Calculation of the activation energy by PS method is shown in **Figure** 7. The expression deduced by Chen [13] and valid for any kinetics is given by Eq. (9), where α stands for τ, δ, and ω, in which the low-temperature half width $\tau = T_m - T_1$, the high-temperature half width $\delta = T_2 - T_m$, and the total half intensity width $\omega = T_2 - T_1$. The values of c_α and b_α are summarized as defined in Eq. (6).

$$E_\alpha = c_\alpha\left(\frac{kT_m^{\,2}}{\alpha}\right) - b_\alpha(2kT_m) \tag{5}$$

$$\begin{cases} c_\tau = 1,51 + 3,0\left(\mu_g - 0,42\right) \\ b_\tau = 1,51 + 4,2\left(\mu_g - 0,42\right) \\ c_\delta = 0{,}976 + 7,3\left(\mu_g - 0,42\right) \\ b_\delta = 0 \\ c_\omega = 2,52 + 10,2\left(\mu_g - 0,42\right) \\ b_\omega = 1 \end{cases} \tag{6}$$

where μ_g is the so-called geometrical shape or symmetry factor that determines the order of the kinetics. The order of the kinetics depends on the glow PS. The value of μ_g for first- and second-order kinetics is 0.42 and 0.52, respectively. The symmetry factor in GOK model can be evaluated from the following Eq. (7). The TL glow peaks corresponding to second-order kinetics are characterized by an almost symmetrical shape, whereas first-order peaks are asymmetrical [6].

$$\mu_g = \frac{\delta}{\omega} = (T_2 - T_m)/(T_2 - T_1) \tag{7}$$

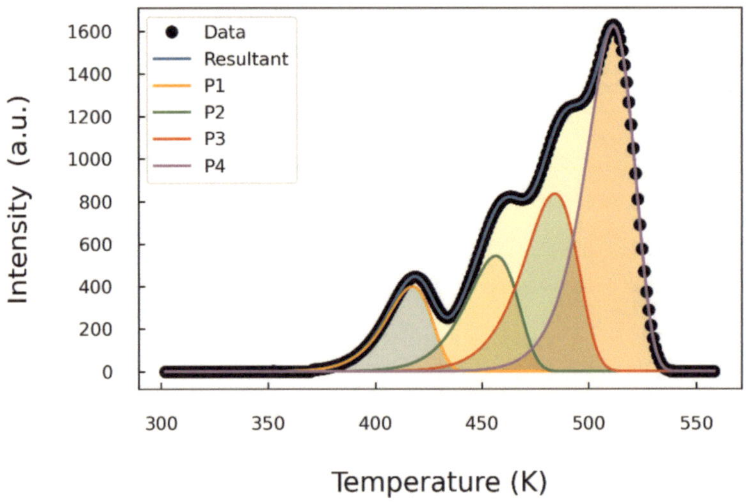

Figure 8.
The TL spectrum of the R4 sample is matched with four peaks.

Explanation of **Figure 7**: describes how to fit the Gaussian peak spectrum and the PS method to calculate the trap energy. The calculation results depend on the T_1, T_2, and T values of the TL peak.

Code in Python:

```
def AE_gen(x,a,b,c,bv):
    y=gaussian(x,a,b,c,bv)
    Tm = geneticTL(x,a,b,c,bv)
    Im = geneticTLy(x,a,b,c,bv)
    y_half=np.zeros_like(x)+Im/2
    print("I=",Im)
    idx = np.argwhere(np.diff(np.sign(y -y_half))).flatten()
    T1=x[idx][0]
    T2=x[idx][1]
    print("Tm=",Tm)
    E=E1(Tm,T1,T2)
    print("E=",E)
    return E
```

TL curves from R package tgcd [14] with 22 TL curves are measured from different materials provided by George Kitis. Among them is the TL curve denoted as R4, which is used to fit the Gaussian peak shape and calculate the kinematic parameters. The results are shown in **Figure 8**.

Details and full Python source code for calculating the activation energy can be found on GitHub [5].

Explanation of **Figure 8**: spectrum of R4 consisting of four peaks matched and peaked by genetic algorithm. The kinetic parameters of the spectrum are calculated reasonably, and the obtained FOM coefficient is very small.

2.4 Calculation of the frequency factor

The intensity I(t) of the TL signal is measured at time t after the start of the experiment. This magnitude TL I(t) is proportional to the derivative −dn/dt,

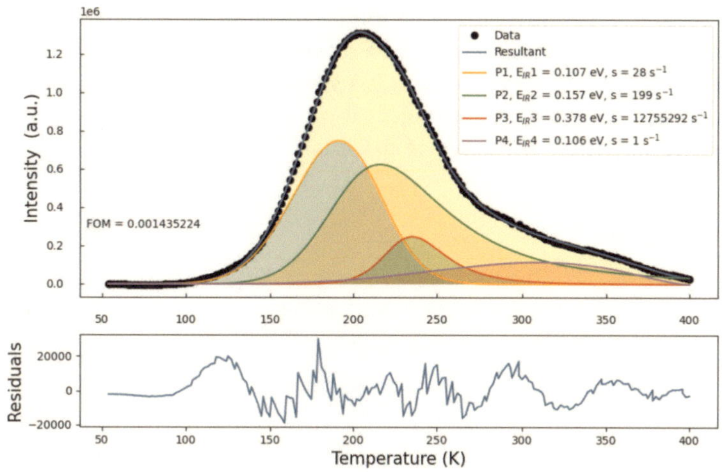

Figure 9.
The spectrum of K2GdF5:Tb curve and results of calculating E and s values.

depending on the measurement conditions. In experimental research, experimentalists pay much attention to the frequency factor of the TL signal. Kitis et al. [7] obtain the following analytic equation for s with the GOK model:

$$s = \left(\frac{\beta E}{k{T_m}^2}\right) \exp\left(\frac{E}{kT_m}\right) \left[1 + (b-1)\frac{2kT_m}{E}\right]^{-1} \tag{8}$$

Thus, kinematic parameters such as E and s of the TL curve will be estimated according to Eqs. (5) and (8). Each peak coordinate of the TL curve including two parameters of temperature and TL intensity will be recorded after each mouse click on the screen containing the TL curve. The kinematic order of the GOK model is also selected and changed from 1 to 2 until the FOM values of the TL curve reach the minimum condition.

Code in Python:

```
def ffGOK(x,a,b,c,bv):
    kbz = 8.617385e-5
    hr=1
    return (hr*c)/(kbz*b**2)
    *np.exp(c/kbz/b)/(1+(bv-1)*1*kbz*b/c)
```

K2GdF5 materials doped with concentrations of 10%Tb are widely used in radiation dosimetry and materials science [15]. The spectrum of K2GdF5:Tb curve and results of calculating E and s values are shown in **Figure 9**.

Explanation of **Figure 9**: depicts the result of spectral matching of sample K2GdF5:Tb with four peaks. The calculation results of sample K2GdF5:Tb in terms of E and s values are also calculated with high accuracy.

Details and full Python source code for calculating the frequency factor can be found on GitHub [5].

3. Conclusion

The application of GA to TL curve analysis provides an efficient and effective method for the estimation of kinetic parameters such as activation energy, frequency factor, and order of reaction. The algorithm can explore a large search space of candidate solutions and converge toward a solution that optimizes the fitness function. The use of GA in TL curve analysis has significant implications for the field of geochronology and archeological dating, as it provides a powerful tool for the accurate and precise determination of the age of materials.

Author details

Nguyen Duy Sang
Can Tho University, Can Tho, Vietnam

*Address all correspondence to: ndsang@ctu.edu.vn

References

[1] Sang ND, Hung NV, Hung TV, Hien NQ. Using the computerized glow curve deconvolution method and the R package tgcd to determination of thermoluminescence kinetic parameters of chilli powder samples by GOK model and OTOR one. Journal of Nuclear Instruments Methods B. 2017;**394**:113-120

[2] Sang ND. Estimate half-life of thermoluminescence signals according to the different models by using python. Journal of Taibah University for Science. 2021;**15**(1):599-608

[3] Sang ND, Thi HHQ. Using the genetic algorithm to detect kinetic parameters of thermoluminescence glow curves. Nuclear Instruments and Methods in Physics Research Section B: Beam Interactions with Materials and Atoms. 2022;**517**:33-42

[4] Coley DA. An Introduction to Genetic Algorithms for Scientists and Engineers. World Scientific; 1999. p. 244

[5] Sang ND. Github. 2023. Available from: https://github.com/sangduynguyen/ga_book

[6] Pagonis V, Kitis G, Furetta C. Numerical and Practical Exercises in Thermoluminescence. United States of America: Springer; 2006

[7] Kitis G, Gomez-Ros JM, Tuyn JWN. Thermoluminescence glow-curve deconvolution functions for first, second and general orders of kinetics. Journal of Physics D: Applied Physics. 1998;**31**(19): 2636-2641

[8] Garlick GFJ, Gibson AF. The electron trap mechanism of luminescence in Sulphide and silicate phosphors. Proceedings of the Physical Society. 1948;**60**:574-590

[9] Sang ND. Study of the effect of gamma-irradiation on the activation energy value from the thermoluminescence glow curve. Journal of Taibah University for Science. 2017; **11**(6):1221-1221

[10] Rawat NS et al. Use of initial rise method to analyze a general-order kinetic thermoluminescence glow curve. Nuclear Instruments and Methods in Physics Research Section B: Beam Interactions with Materials and Atoms. 2009;**267**(20):3475-3479

[11] Correcher V, Garcia-Guinea J. Potential use of the activation energy value calculated from the thermoluminescence glow curves to detect irradiated food. Journal of Radioanalytical and Nuclear Chemistry. 2013;**298**(2):821-825

[12] McKeever SWS. Thermoluminescence of Solids. Cambridge: Cambridge University Press; 1988

[13] Chen R. On the calculation of activation energies and frequency factors from glow curves. Journal of Applied Physics. 1969;**40**:570

[14] Peng J, Dong Z, Han F. Tgcd: An R package for analyzing thermoluminescence glow curves. SoftwareX. 2016;**5**:112-120

[15] Van Tuyen H et al. Spectroscopic studies of K2GdF5:Nd3+ single crystals for incredibly strong NIR emission at 864 nm. Journal of Physics and Chemistry of Solids. 2022;**161**:110454